INTERNATIONAL CENTRE FOR MECHANICAL SCIENCES

COURSES AND LECTURES   No. 237

STRUCTURAL OPTMIZATION

EDITED BY

P. BROUSSE

UNIVERSITE' DE PARIS VI

SPRINGER-VERLAG WIEN GMBH

ISBN 978-3-211-81376-8   ISBN 978-3-7091-2738-4 (eBook)

DOI 10.1007/978-3-7091-2738-4

PREFACE

In the last decades, new methods of analysis and the advent of computers have permitted a large development of the theory and the computation of mechanical structures. Further, the need of economy and efficiency has led scientists and engineers to look for more improved structures and to optimize them in view of a specific use. For example, among the simplest problems posed by industry are the following: in astronautics, the minimization of the mass so as to reduce running expenses; in general engineering, the maximization of the supported loads, or the maximization of the fundamental frequency of vibrations in order to reduce the risk of resonance; in precision industry, the minimization of the dimensions or the optimization of the form, etc.

In general, concrete problems of structural optimization imply great difficulties. First, they often require hard work for the choice of the function to be minimized or maximized as well as for the choice of the design variables on which it is possible and reasonable to act. Secondly, even after these problems are resolved in distinct terms, serious difficulties still persist: it is necessary to analyse the mechanical structure; new theoretical optimization problems can appear; often algorithms and programmes must be developed; it may be necessary to have powerful computers, etc.

The needs of industry have engendered numerous studies in the abovementioned fields: theory of structures, constitutive laws, schematization, systematic use of numerical methods. On the other hand, new progress has been made in mathematical optimization, in variational methods, in non-convex problems, in numerical methods of optimization, in the extension of the validity of results already known, etc. These studies have been sponsored by national authorities of several countries, by industrial bodies and by private firms. Numerous papers have been published on these subjects in various scientific and technical journals.

In view of this large scope of research the International Centre for Mechanical Sciences considered that it was of interest to organize basic courses on problems and methods in the optimization of mechanical structures. The purpose of these courses was to convey to the participants a good understanding of what was already done, and, at the same time, to stimulate them in creating new techniques and solutions. The aim was therefore to provide general methods of optimization with application fields as large as possible, to treat completely concrete examples, to point out some open problems, to give future perspectives, etc. The C.I.S.M. initiative was successful and the lectures, the discussions and the seminars were attended with assiduity and enthusiasm by numerous researchers.

The lecturers and animators of these courses were P. Brousse, A. Čyras, G. Maier, W. Prager and M.A. Save. Prager's course was already published and Maier's course was based on several papers already published. In this book, we present the courses of P. Brousse, A. Čyras and M.A. Save. Here, we give a brief summary of the subjects.

The first part (P. Brousse) deals with general optimization methods. The purpose is to present basic methods and theorems in a manner as complete as possible, and to show, by means of examples, how they can be used. The topics discussed are: linear optimization, the simplest properties about convexity, unconstrained problems, constrained problems, Lagrange and Kuhn-Tucker methods, saddle-point properties, duality, some numerical methods, some examples of optimization of statical and vibrating structures formed of linear elements.

The second part (A. Čyras) is concerned with the optimization theory in the design of elastic-plastic structures. A generalized formulation of some problems of limit analysis is given on the ground of the basic optimization theorems. Two optimality criteria are considered: the first one consists in maximizing a function related to the loads, whereas the second one consists of minimizing a function related to the qualities of the body. A statical and a kinematical formulations are presented and explained. The Lagrange method and the duality are widely used. Increasing loading and cyclic loading are successively considered, and the results obtained in these two types of loadings are compared with each other.

The third part (M.A. Save) deals with a review of the current knowledge in the optimal plastic design of structures. A theory is given at first in a simple case which leads to an optimality criterion. This is then successively generalized to moving loads, stepwise varying design unknown, action of body-forces and general convex cost functions. At last, the author presents his general criterion for optimal structural design.

In each chapter, the theorems are proved and numerous examples with numerical results are provided.

I cannot end without expressing my thanks to the Scientific Council of the C.I.S.M., to the Administrative Council and to Rector W. Olszak for their intrusting me with the organisation of these courses. My colleagues and myself have found great satisfaction in the presentation of the courses and in the discussions which followed. We would also like to thank the printing office of the C.I.S.M. and the Springer Verlag for the perfection with which they have assured the publishing and the distribution of this book.

P. Brousse

CONTENTS

Chapter II. Nonlinear Optimization

A – Convexity

B – Definitions and Basic Properties of Optimization without Differentiability

C – Unconstrained Optimization *Problems*

D – Optimization with Equality Constraints

E – Optimization with Inequality Constraints. Duality

F – Applications to Mechanical Problems

Chapter III. Some Numerical Methods
A — Problems without Constraints

B — Problems with Constraints

Optimization Theory in the Design of Elastic-Plastic Structures
by A. Cyras

Chapter I. General Statements

Chapter II. An Elastic-plastic Body Under Monotonically Increasing Loading

Chapter IV. Second Generalization: Stepwise Varying Design Unknown

Chapter V. Third Generalization: Action of Body-Forces

Chapter VI. Fourth Generalization: Convex Cost Function

Chapter VII. General Discussion of Plastic Optimal Design   195

Chapter VIII. A General Criterion for Optimal Structural Design

GENERAL METHODS OF OPTIMIZATION IN STRUCTURAL MECHANICS

by

Pierre BROUSSE
Université de Paris VI

CHAPTER I

LINEAR OPTIMIZATION[(i)]

Numerous problems of structural optimization are linear or can reasonably be approached by linear problems. This is why linear optimization is very important.

In part A, we present the various formulations met in such problems, and we introduce some notions conducting to the simplex method. Part B deals with duality in linear optimization. Finally, in part C, we illustrate the results obtained by an example relevant to Civil Engineering: the minimization of the mass of rigid plastic trusses.

We use the following notations; the data are

$$
\begin{array}{ll}
A = (a_{ij}) & : \quad (m,n) - matrix \\
b = \{b_i\} & : \quad non\ null\ (m,1) - matrix\ (m-vector) \\
c = \{c_j\} & : \quad non\ null\ (n,1) - matrix\ (n-vector)
\end{array}
\tag{1}
$$

The unknown quantities are

$$
\begin{array}{ll}
x = \{x_j\} & : \quad n-vector \\
y = \{y_i\} & : \quad m-vector
\end{array}
\tag{2}
$$

A. GENERAL PROPERTIES

I-1 Various Formulations of Linear Optimization

I-1.1 <u>Statements.</u> Because of the concrete origin of practical problems and for the sake of convenience, we replace every numerical variable which is restricted in sign by two positive[(ii)] variables: $u = u^+ - u^-$ with $u^+ \geqslant 0$ and $u^- \geqslant 0$. Then, there is no loss of generality in supposing that the variable x is positive: $x_j \geqslant 0$ for all $j = 1,...n$.

(i) The basic results on linear optimization are, from the outset, necessary to other courses. It is the reason why we treat this question separately in an opening chapter.

(ii) The word "positive" (resp. "negative") will always mean $\geqslant 0$ (resp. $\leqslant 0$).

Now, the formulation of linear optimization with linear inequality constraints is

(3)
$$\text{Problem P} \begin{cases} \text{minimize } c^T x \\ \text{on } F : Ax \geq b \; ; \; x \geq 0 \end{cases}$$

The formulation with linear equality constraints is

(4)
$$\text{Problem P}_1 \begin{cases} \text{minimize } c^T x \\ \text{on } F_1 : \; Ax = b \; ; \; x \geq 0 \end{cases}$$

The mixed formulation with equality constraints and inequality constraints is

(5)
$$\text{Problem P'} \begin{cases} \text{minimize } c^T x \\ \text{on } F' : A'x \geq b' \, , \, A''x = b'' \; ; \; x \geq 0 \end{cases}$$

where A' and A'' (resp. b' and b'') are two submatrices of A (resp. b) defined by

$$A = \begin{pmatrix} A' \\ A'' \end{pmatrix} \quad , \quad b = \begin{pmatrix} b' \\ b'' \end{pmatrix}$$

A', b' having m' rows and A'', b'' having $m - m' = m''$ rows.

In each of the three formulations P, P_1 , P' the function $c^T x$ is the <u>cost</u> or the <u>objective function</u>; the subset F (or F_1, or F') of R^n is called the <u>feasible region</u>.

I-1.2 <u>Transformations</u>. In fact, each of these formulations can be transformed into the two others. For instance, let us see how we can go from P' to P_1 .

A slack variable is added to each inequality constraint in such a way that it is converted into an equality. If $w_{j'}$, $j' = 1, \dots m'$, are these variables, we write $w = \{w_{j'}\}$ and we call $I_{m'}$, the unit (m', m')–matrix. Now, P' can be written

(6)
$$\begin{cases} \text{minimize } (c^T \; o^T) \begin{pmatrix} x \\ w \end{pmatrix} \\ \text{on } \begin{pmatrix} A' & -I_{m'} \\ A'' & \square \end{pmatrix} \begin{pmatrix} x \\ w \end{pmatrix} = b \; ; \; \begin{pmatrix} x \\ w \end{pmatrix} \geq 0 \end{cases}$$

where 0 (resp. $\square$) is the null (m', 1) (resp. (m", m'))—matrix. It is really a P_1 formulation for the variable ($\begin{smallmatrix} x \\ w \end{smallmatrix}$).

I—2 Some Properties and General Assumptions

I-2.1 <u>Set of Solutions</u>. In all three cases P, P_1, P', the feasible region F is convex and closed, because it is the intersection of planes or closed half-spaces: it is bounded from below (cf. $x \geqslant 0$). The function $c^T x$ can be minimum at several points of F. but its minimum (when it exists) is unique.

From the last remarks, we deduce the following properties:

$\left\{ \begin{array}{l} \underline{\text{If the problem P (resp. } P_1 \text{, P') has a solution } x^o \text{, then the set of solutionses the}} \\ \underline{\text{intersection of F (resp. } F_1 \text{, F') and of the plane } c^T x = c^T x^o. \text{ This set is convex,}} \\ \underline{\text{closed and bounded from below.}} \end{array} \right.$ (7)

For every point x of F, or F_1, or F' we have $c^T x^o \leqslant c^T x$. Thus. the plane $c^T x = c^T x^o$ is a supporting plane at x^o.

I-2.2 <u>Rank of the Matrices of Problem P_1</u> . Thus far, we have made no assumption neither on the numbers m and n nor on the rank r of the matrices defined above.

— Let us write (A b) the matrix formed by justaposing the matrices A and b. We always have $r(A) \leqslant r$ (A b). If $r(A) < r(A b)$, the equation Ax = b would have no solution and problem P_1 would not have any meaning. <u>In problem P_1 we shall suppose that $r(A) = r(A b)$</u>. Let k be this common rank.

— We always have $k \leqslant m$. Let us examine the case $k < m$. In the matrix (A b), there would exist k linearly independent rows, and the (m−k) others would be linear combinations of these k ones. The (m − k) corresponding equations are redundant and can be cut out. <u>In problem P_1 we shall suppose that $r(A) = m$.</u>

— We always have $r(A) \leqslant n$, that is to say (according to the latter assumption) $m \leqslant n$. If m were equal to n, then matrix A would be a regular matrix, the equation Ax = b would have a unique solution noted x^1, so that problem P_1 would have no solution if $x^1 < 0$ or a unique solution if $x \geqslant 0$. <u>In problem P_1 we shall suppose that $m < n$</u> (number of equations strictly inferior to the number of variables).

I–3 Extreme Points and Vertices of the Feasible Region F_1 of Problem P_1

I-3.1 <u>Definitions</u>. A point x of a convex subset C of an affine space is said to be an extreme point of C, if and only if, there do not exist in C two distinct points symmetrical with respect to x.

The extreme points of the polyhedron F, or F_1, or F' (problems P, P_1, P') are also called vertices.

We are going to characterize the vertices of the feasible region F_1 of problem P_1.

Let us remark that, according to $b \neq 0$, F_1 does not contain the origin point. Therefore, <u>every point of F_1 has at least one strictly positive component</u>.

Now, the following theorem holds:

(8) $\left\{ \begin{array}{l} \underline{\text{In order that a point } \bar{x} \text{ of } F_1 \; (Ax = b, \, x \geqslant 0) \text{ be a vertex, it is necessary}} \\ \underline{\text{and sufficient that the columns of A corresponding to the strictly}} \\ \underline{\text{positive components of } \bar{x} \text{ be linearly independent.}} \end{array} \right.$

<u>Proof</u>. Let $I(\bar{x})$ be the subset made of the j's such that $\bar{x}_j > 0$, and let a^j be the jth column of A. We have

(9) $$\sum \bar{x}_j a^j = b \quad , \quad j \in I(\bar{x}) .$$

— First, let us suppose that $\bar{x}$ is a vertex of F_1. We have to prove that the columns a^j, $j \in I(\bar{x})$, are linearly independent.

Should it be otherwise, there would exist numbers λ_j, not all of them null, such that

(10) $$\sum \lambda_j a^j = 0 \quad , \quad j \in I(\bar{x}) .$$

Let λ be the vector of R^n having as components the λ_j's if $j \in I(\bar{x})$ and 0 if $j \notin I(\bar{x})$. The vector λ is not null. Because of (9) and (10), the straight line

(11) $$x = \bar{x} + \rho\lambda \quad , \quad \rho \text{ arbitrary number,}$$

is contained in Ax = b. On this straight line, the segment defined by

$$|\rho| \leqslant \min \frac{\overline{x}_j}{|\lambda_j|} \, , \, j \in I(\overline{x}), \, \lambda_j \neq 0 \tag{12}$$

verifies $x \geqslant 0$. Since the right-hand side of (12) is strictly positive, there exist in F_1 some distinct points symmetrical with respect to $\overline{x}$. Therefore, $\overline{x}$ is not a vertex. This is in contradiction with the initial assumption. So, the columns a^j, $j \in I(\overline{x})$, are linearly independent.

— Conversely, suppose that the columns a^j, $j \in I(\overline{x})$, are linearly independent. Let x', x'' be two points of F_1 symmetrical with respect to $\overline{x}$. We have successively

$$x'_j = x''_j = 0 \quad \text{if} \quad j \notin I(\overline{x}) \tag{13_j}$$

$$\Sigma \, x'_j a^j = b \, \text{ and } \, \Sigma \, x''_j a^j = b \, , \, j \in I(\overline{x})$$

$$\Sigma \, (x'_j - x''_j) a^j = 0 \, , \, j \in I(\overline{x})$$

$$x'_j = x''_j \quad \text{if} \quad j \in I(\overline{x}) \, . \tag{13_j}$$

The equalities (13_j) show that $x' = x'' = \overline{x}$. The point $\overline{x}$ is a vertex of F_1.

I-3.2 <u>Degeneracy</u>. <u>Basic Variables</u>. We recall the assumptions made on F_1: $m < n$, $r(A) = m$.

According to theorem (8), a vertex of F_1 has at most m non null components. It is said to be non degenerated (resp. degenerated) if and only if the number of its non null components is equal (resp. strictly inferior) to m.

— Let $\overline{x}$ be a vertex of F_1. In A, the columns numbered j, $j \in I(\overline{x})$, are linearly independent. They form an (m, h)—matrix, called $M(\overline{x})$; the rank h of $M(\overline{x})$ verifies $h \leqslant m$. If $\overline{x}$ is non degenerated, then $M(\overline{x})$ is a regular square matrix. If $\overline{x}$ is degenerated, it is always possible to complete $M(\overline{x})$ by $(m-h)$ other columns of A in such a way that the completed matrix, called $B(\overline{x})$, is regular. Besides, the matrix $B(\overline{x})$ is not always unique.

— Conversely, let B be a regular (m, m)—matrix extracted from A. There is no loss of generality by supposing that B is formed by the first m columns of A. Then, in $Ax = b$, there exists a point $\overline{x}$ whose first m components form the matrix $\overline{x}_B = B^{-1} b$ and whose $(n-m)$ others are 0. If $\overline{x}_B \geqslant 0$, then $\overline{x}$ is a

vertex (degenerated or not).
— Every regular (m, m) matrix extracted from A is called a base. The variables
   or components associated with the columns of a base B are said to be <u>basic</u>
   <u>variables</u> or <u>basic components.</u> All the non-basic components of a vertex of
   F_1 are null.

I−4 Vertex Solutions (Basic Solutions) of Problem P_1

Every vertex solution of problem P_1 is also called a <u>basic solution</u>. About these
solutions, the following fundamental <u>theorem</u> holds.

(14) | If problem P_1 has a solution, it has at least a basic solution

<u>Proof.</u> Let $\bar{x}$ be the given solution. If the columns a^j of A, $j \in I(\bar{x})$, are
linearly independent, then $\bar{x}$ is a vertex, and the theorem is proved. Otherwise, we
consider die segment defined by (11) and (12). We have noted that this
segment is in F_1. It does not pass through the supporting plane $c^T x = c^T \bar{x}$.
Therefore it lies in this plane, and all its points are solutions of problem P_1. It is the
case of the point x* obtained when ρ verifies <u>equality</u> (12), for some value of j,
noted j*. The j^{*th} component of x* is null. Then, the number of non null
components of x* is strictly smaller than the number of non null components of $\bar{x}$.
If x* is a vertex, the theorem is proved. Otherwise, we have to repeat the previous
construction a certain number of times. A vertex solution is always obtained,
because all the points of F_1 have at least a non-null component (b $\neq$ 0).

I−5 Aim of the Simplex Method

The number of vertices of F_1 is finite. According to theorem (14), in order to
treat problem P_1 numerically, it would be sufficient to compute the value of the
cost at all vertices. It would be necessary to compute all the vertices. But the
number of vertices can be equal to the number of combinations m to m of the n
columns of A, that is to say n! /m! (n − m)! . This number can become very large,
so that even the leading computers are incapable of such a performance.

The simplex algorithm consists in
— determining an initial vertex,
— recognizing if this vertex is a solution of problem P_1 or not (sometimes, the
   same test permits to recognize some cases where the problem has no

solution),

— possibly, choosing another easily computable vertex, implying reduction of the cost.

Then, from vertex to vertex suitably chosen, one succeeds either in showing that the problem has no solution or in computing all the vertex solutions.

Now, it should be necessary to make a remark relative to the assumption $r(A) = m$.

When m is large, it is extremely difficult to know whether the rows of A are linearly independent or not. Then, some modified constraints are used often introducing new variables, in such a way that the rank of the new matrix obtained in this manner is equal to m. It happens that the simplex method enables us to know whether or not there were in the first system $Ax = b$ some redundant equations, and in the affirmative case to take away a system of redundant equations. Thus, in practice, the assumption $r(A) = m$ is not restrictive.

B. EXISTENCE OF SOLUTIONS AND DUALITY

I-6 Farkas Theorem

In order not to break off the sequence of the statements of this subchapter, we give, first, an important preliminary theorem.

$$\left\{ \begin{array}{l} \underline{\text{Farkas theorem.}} \ \underline{\text{Let M be a given}} \ (q, n)-\underline{\text{matrix and let}} \ c \ \underline{\text{be a}} \\ \underline{\text{given n-vector. In order that all the solutions of the inequation}} \\ Mx \geqslant 0 \ \underline{\text{verify}} \ c^T x \geqslant 0, \ \underline{\text{it is necessary and sufficient that there}} \\ \underline{\text{exists a q-vector}} \ \lambda \geqslant 0 \ \underline{\text{such that}} \ M^T \lambda = c. \end{array} \right. \tag{15}$$

Proof of the Sufficiency. Suppose that x verifies $Mx \geqslant 0$ and that there exists a q-vector $\lambda \geqslant 0$ such that $M^T \lambda = c$. We have successively

$$c^T x = \left(M^T \lambda \right)^T x = \lambda^T (Mx) \tag{16}$$

The two factors λ^T and Mx are positive; so is $c^T x$.

Proof of the Necessity. We are going to prove the following property which implies the necessity.

(17) $\left\{ \begin{array}{l} \text{If there exists no q-vector } \lambda \text{ verifying simultaneously in-} \\ \text{equality } \lambda \geqslant 0 \text{ and equality } M^T \lambda = c, \text{ then there exists an} \\ \text{n-vector } x^o \text{ such that } Mx^o \geqslant 0 \text{ and } c^T x^o < 0. \end{array} \right.$

Let C be the cone generated in R^n by the q columns of M^T, that is to say the set of the points z of R^n such that there exists a q-vector, μ, verifying $\mu \geqslant 0$ and $M^T \mu = z$. This cone is convex and closed.

The assumption made in (17) means that the point c is not in C. Therefore, the set C and the point c have a strictly separating plane: there exist a non null n-vector x^o and a number α such that

$$(18) \qquad\qquad\qquad\qquad (x^o)^T c < \alpha$$

and

$$(19) \qquad\qquad (x^o)^T v > \alpha \quad \text{for all vectors (points) v of C .}$$

i) We have $\alpha < 0$. Indeed, the inequality (19) holds for all points of C, in particular for $v = 0$: the number α is strictly negative.

Then, according to (18), we have

$$(20) \qquad\qquad\qquad\qquad c^T x^o < 0$$

ii) <u>All vectors v of the cone</u> C verify $v^T x^o \geqslant 0$. Let us suppose that there exists a vector v^o of C such that $(v^o)^T x^o < 0$. By taking $k > \alpha/(v^o)^T x^o$, we have $(x^o)^T (kv^o) < 0$, which contradicts inequality (19) written for the vector kv^o.

Then all vectors v of C verify $v^T x^o \geqslant 0$.

iii) In particular, all the column-vectors of M^T verify this last inequality. Therefore we have

$$M^T x^o \geqslant 0 \qquad (21)$$

From (20) and (21) we deduce (17).

I-7 Kuhn-Tucker Theorems

First, let us introduce a new problem and a new notation.

> New problem. Minimize $c^T x$ on ϵ : $Ax \geqslant b$.   (22)

> New notation. Given an arbitrary point $\bar{x}$ of ϵ we shall call
> $N(\bar{x})$ the subset of all the numbers j of (1, ...m) such that the   (23)
> j^{th} constraint is "active" at $\bar{x}$, i.e. $a_j \bar{x} = b_j$ (where a_j is the
> j^{th} row of A).

Now, we prove a lemma.

I-7.1

> Lemma. A point $\bar{x}$ contained in ϵ is a solution of problem (22)
> if and only if, for each n-vector h verifying $a_j h \geqslant 0$ for all j   (24)
> in $N(\bar{x})$, we have $c^T h \geqslant 0$.

Proof. According to the definition of the minimum, a point $\bar{x}$ of ϵ is a solution of problem (22) if and only if

$$c^T h \geqslant 0 \qquad (25)$$

for all n-vectors h such that the point $\bar{x} + h$ is contained in ϵ .

Necessity. The proof of the necessity uses the following remark: if a vector h verifies $a_j h \geqslant 0$ for all j's in $N(\bar{x})$, then there exists a number $\rho > 0$ such that the point $\bar{x} + \rho h$ is in ϵ (note that for $j' \notin N(\bar{x})$, we have $b_{j'} - a_{j'} \bar{x} < 0$; take $\rho > 0$ such that $\rho (a_{j'} h) \geqslant b_{j'} - a_{j'} \bar{x}$ for every $j' \notin N(\bar{x})$).

Now, let $\bar{x}$ be a solution of problem (22), and let h be an n-vector such that $a_j h \geqslant 0$ for all j's in $N(\bar{x})$. Then, for each number ρ previously determined, the point $\bar{x} + \rho h$ is contained in ϵ . Therefore, according to (25), we have $c^T(\rho h) \geqslant 0$ and consequently $c^T h \geqslant 0$.

Sufficiency. Let h be an arbitrary n-vector such that the point $\bar{x} + h$ is in ϵ .

Then, for all j's in $N(\bar{x})$, we have $a_j(\bar{x} + h) \geqslant b_j$ and therefore $a_j h \geqslant 0$. But, from the hypothesis of (24), we have $c^T h \geqslant 0$. Then, $\bar{x}$ is solution of problem (22).

I-7.2

(26) $\left\{ \begin{array}{l} \underline{\text{Kuhn-Tucker theorem for problem}} \text{ (22)}. \underline{\text{A point}} \bar{x} \underline{\text{of}} \epsilon \underline{\text{is a}} \\ \underline{\text{solution of problem}} \text{ (22)} \underline{\text{if and only if there exists an m-vector}} \\ \text{u} \underline{\text{such that}} \\ \qquad u \geqslant 0 \ , \ A^T u = c \ , \ u^T(A\bar{x} - b) = 0 \end{array} \right.$

Proof. If the point $\bar{x}$ is a solution of (22), it is on the boundary of ϵ (see I-2.1). Therefore $N(\bar{x})$ is non-empty.

Now, let us consider an arbitrary point $\bar{x}$ of the boundary of ϵ . There is no loss of generality by supposing $N(\bar{x}) = \{1, ... q\}$. Then, let M be the matrix formed for the first q rows of A. According to the previous lemma (24), $\bar{x}$ is a solution of (22) if and only if, for all vectors h verifying $Mh \geqslant 0$, we have $c^T h \geqslant 0$, that is to say (by Farkas theorem) if and only if there exists a q-vector $\lambda \geqslant 0$ such that $M^T \lambda = c$.

i) If x is a solution of (22), we complete the vector λ of R^q to a vector u of R^m by adding $(m - q)$ zero components. Then we have $A^T u = c$. On the other hand, for j in $N(\bar{x})$ (resp. not in $N(\bar{x})$)) we have $a_j \bar{x} - b_j = 0$ (resp. $u_j = 0$). The necessity is proved.

ii) Conversely, suppose that the three relations of (26) are satisfied. According to the third one, if j is not in $N(\bar{x})$, then $u_j = 0$. Now, let us delete in A the last $(m - q)$ rows and let M be the matrix obtained in this way. Likewise, let λ be the q-vector formed by the first q components of u. Then, equality $A^T u = c$ becomes $M^T \lambda = c$. The sufficiency is proved.

I-7.3

(27) $\left\{ \begin{array}{l} \underline{\text{Kuhn-Tucker theorem for problem}} \text{ P (cf. (3))}. \underline{\text{A point}} \bar{x} \underline{\text{of}} \text{ F} \underline{\text{is a}} \\ \underline{\text{solution of problem}} \text{ P} \underline{\text{if and only if there exists an m-vector}} \bar{y} \\ \underline{\text{such that}} \\ \qquad \bar{y} \geqslant 0 \quad , \qquad A^T \bar{y} \leqslant c \\ \underline{\text{and moreover verifying either}} \\ \qquad \bar{y}^T (A\bar{x} - b) = 0 \quad \text{and} \quad \bar{x}^T (c - A^T \bar{y}) = 0 \\ \underline{\text{or}} \\ \qquad c^T \bar{x} = b^T \bar{y} \end{array} \right.$

<u>Proof</u>. We define F by

$$\begin{pmatrix} A \\ I_n \end{pmatrix} x \geqslant \begin{pmatrix} b \\ 0 \end{pmatrix}$$

where I_n is the (n, n)—unit matrix and 0 the null n-vector. According to (26), a point $\bar{x}$ of F is a solution of problem P if and only if there exists an m-vector $\bar{y}$ and an n-vector $\bar{z}$ such that

$$\begin{cases} \bar{y} \geqslant 0 \quad , \quad \bar{z} \geqslant 0 \\ A^T \bar{y} + \bar{z} = c \\ \bar{y}^T (A\bar{x} - b) + \bar{z}^T \bar{x} = 0 \end{cases}$$

that is to say, by eliminating $\bar{z}$, either

$$\bar{y} \geqslant 0 \quad , \quad A^T \bar{y} \leqslant c \quad , \quad \bar{y}^T (A\bar{x} - b) + \bar{x}^T (c - A^T \bar{y}) = 0 \qquad (28)$$

$$\text{or} \qquad \bar{y} \geqslant 0 \, , \, A^T \bar{y} \leqslant c \, , \, c^T \bar{x} = b^T \bar{y} . \qquad (29)$$

In the last of the relations (28), the two terms are positive. Their sum is null if and only if they are both null.

Then, the Kuhn-Tucker theorem for problem P is proved.

I-8 Definition of Duality

First, we give the definition of the dual of problem P. From this, we shall deduce the definition in the general case of the dual problem P' (and consequently of problem P_1).

I-8.1 Problem P

The dual of problem P is problem D stated below.

$$\begin{cases} \underline{\text{Problem P}} - \underline{\text{minimize}} \ c^T x \quad \underline{\text{on}} \ F_P^{(i)} : Ax \geqslant b, \ x \geqslant 0 \\ \underline{\text{Problem D}} - \underline{\text{maximize}} \ b^T y \quad \underline{\text{on}} \ F_D : A^T y \leqslant c, \ y \geqslant 0 \end{cases} \qquad (30)$$

The new unknown y is an m-vector; it is called the dual variable.

Problem D is equivalent to:

$$\text{minimize } (-b)^T y \text{ on } F_D : (-A^T) y \geqslant (-c), \ y \geqslant 0$$

_____

(*) Now, we write F_P (primal feasible region) instead of F.

According to definition (30), its dual is:

maximize $(-c)^T x$ on $(-A)x \leqslant (-b)$, $x \geqslant 0$.

This is problem P. Then the following property holds.

(31) | Problems P and D are dual of each other.

It should be remarked that the duality introduces a correspondence between the jth variable x_j of P and the jth constraint of D: $\sum_{i=1}^{m} a_{ij} y_i \leqslant c_j$. By the reciprocity of the duality we obtain

(32) | The jth variable of each problem P, D corresponds to the jth constraint of the other and conversely.

I-8.2 Problem P'

First, we write P' in the form of a problem with inequalities only (cf. I-1.2). Afterwards, we introduce the dual variable.

$$\eta = \begin{pmatrix} y' \\ z \\ w \end{pmatrix}$$

where y' (resp. z and w) has the same number of rows as A', b' (resp. as A'', b''). Then, according to definition (30), the dual problem of P' is:

$$\text{maximize } b^T y \text{ on } (A'^T \quad A''^T \quad -A''^T)\eta \leqslant c, \eta \geqslant 0$$

We set $y'' = z - w$ and we see that y'' is unrestricted in sign. Conversely, if y'' is given, it can be decomposed into its positive and negative parts. Thus, problem P' and the following problem D' are dual of each other.

(33) | Problem D' - maximize $b^T y$
on F'_D : $A^T y \leqslant c$, $y = \begin{pmatrix} y' \\ y'' \end{pmatrix}$, $y' \geqslant 0$, y'' arbitrary

Then, the rule of formation of the dual problem of P' (and consequently of the dual problem of P and P_1) is as follows

$$\begin{cases} \text{If the } j^{th} \text{ constraint of the primal (problem P') is an inequality,} \\ \text{the } j^{th} \text{ dual variable is positive (i.e.} \geq 0). \text{ If the } j^{th} \text{ constraint of the} \\ \text{primal (problem P') is an equality, the } j^{th} \text{ dual variable is} \\ \text{unrestricted in sign.} \end{cases} \quad (34)$$

I-8.3 <u>Problem P_1</u>. In particular, the next two problems are dual of each other.

$$\begin{cases} \underline{\text{Problem } P_1} - \underline{\text{minimize}} \ c^T x \ \underline{\text{on}} \ F_{1P} \ : Ax = b, \ x \geq 0 \\ \underline{\text{Problem } D_1} - \underline{\text{maximize}} \ b^T y \ \underline{\text{on}} \ F_{1D} : A^T y \leq c \end{cases} \quad (35)$$

I−9 Existence and Duality Theorems

The existence and duality theorems can be deduced from the Kuhn-Tucker theorem (27). They take the directly usable following form.

I-9.1

$$\begin{cases} \text{If the point } \bar{x} \text{ of } F_P \text{ and the point } \bar{y} \text{ of } F_D, \text{ (cf. (30)) verify the} \\ \text{equality } c^T\bar{x} = b^T\bar{y}, \text{ then they are respectively solutions of} \\ \text{problems P and D.} \end{cases} \quad (36)$$

Indeed, according to (27), the point $\bar{x}$ is a solution of P. The reciprocity of duality shows that $\bar{y}$ is a solution of D.

I-9.2

$$\begin{cases} \text{If one of the problems P or D has a solution, so has the other.} \end{cases} \quad (37)$$

Indeed, let us suppose, for example, that x* is solution of problem P. Then, according to (27), there exists in F_D a point $\bar{y}$ such that $c^T x* = b^T\bar{y}$. By (36), $\bar{y}$ is solution of problem D.

I-9.3

(38) $\quad \begin{cases} \text{If } x^* \text{ and } y^* \text{ are respectively solutions of P and D, then} \\ c^T x^* = b^T y^* \end{cases}$

Indeed, property (37) implies that problem D has a solution $\bar{y}$ and that $c^T x^* = b^T \bar{y}$. The minimum value of the cost being unique, we have $c^T x^* = b^T y^*$.

I-9.4

(39) $\quad \begin{cases} \text{If in a solution of one of the two problems P and D, (or P' and} \\ \text{D'), the } j^{th} \text{ constraint is inactive, then, in every solution of the} \\ \text{other, the } j^{th} \text{ variable has a null value.} \\ \text{If in a solution of one of the two problems P and D,} \\ \text{(or P' and D', or } P_1 \text{ and } D_1\text{), the } j^{th} \text{ variable has a non null} \\ \text{value, then in every solution of the other, the } j^{th} \text{ constraint is} \\ \text{active.} \end{cases}$

This property comes from (27) for problems P and D, and from the relations between problems P, P', P_1 (cf. I.1).

C. MINIMIZATION OF THE MASS OF RIGID PLASTIC TRUSSES

We are going to show that the problem of minimization of the mass of rigid plastic trusses leads to linear optimization problems. From theorems proved in subchapters A and B, we shall immediately deduce the main properties of the solutions of this optimal design. Our aim is to show briefly, on this example, how useful in Engineering the linear methods are.

I-10 Stability Conditions

I-10.1 <u>Notations</u>. We consider a general hyperstatic truss. The joints are rigidly fixed, completely free or compelled to stay on given planes or on given straight lines. The virtual velocity of each completely free joint (resp. partially free joint) has three components (resp. two or one component). All these components are arranged in an m-vector, called $\dot{u}$. Let $(\dot{\Delta \ell})_k$ be an arbitrary rate of elongation of the kth bar; these quantities are arranged in a ν-vector $(\dot{\Delta \ell})$. Then, there exists a (ν, m)—matrix, called

E^T , depending only on the geometry of the truss, such that

$$(\dot{\Delta \ell}) = E^T \dot{u} \tag{40}$$

On the completely free joints and on the partially free joints are applied given loads. The components of these loads corresponding to the previous virtual velocity components are set in an m-vector, called p. So that, supposing that the connections are perfect, the virtual power of the given loads is

$$p^T \dot{u} \tag{41}$$

The weight of the bars and of the joints is neglected.

The force carried by the kth bar is called f_k (positive in tension, negative in compression). The f_k 's are set in a ν-vector, f. Then, the power of dissipation of the internal forces is

$$f^T (\dot{\Delta \ell}) \tag{42}$$

I-10.2 <u>Use of the Static Theorem</u>. This theorem can be stated as follows: "the truss can support the given load if and only if there exists an internal force-vector f statically and plastically admissible".

The principle of virtual power gives

$$f^T E^T \dot{u} = p^T \dot{u} \tag{43}$$

for all $\dot{u}$. Then, the statically admissible internal forces are those such that

$$Ef = p \tag{44}$$

In other words, (44) is the <u>equilibrium equation</u>.

Now, we introduce a property of matrix E. Let us consider, instead of the given load-vector p, an arbitrary (virtual) load-vector p* . The equilibrium equation is the same as (44), with the right-hand side replaced by p* . According to the assumption of hyperstaticity, for each p* the new equation has several solutions. Therefore we have

$$m < \nu \quad \text{and} \quad \underline{\text{rank of}} \ E = m \tag{45}$$

We find the hypothesis of (I-2.2) again.

Let us pass on to the plasticity condition. We suppose that the bars are cylindrical and made of a given homogeneous material. Then, each bar is characterized by its cross section area s_k. Let s be the ν-vector $\{s_k\}$. Moreover, we suppose that the yield stress in tension is equal to the yield stress in compression. If σ_o is this yield stress, an internal force-vector f is plastically admissible if and only if it verifies

$$(46) \qquad |f| \leqslant \sigma_o s .$$

Finally, the <u>stability conditions</u> are

$$(47) \qquad \begin{cases} Ef = p \\ |f| \leqslant \sigma_o s \end{cases}$$

I-10.3 <u>Use of the Kinematic Theorem.</u> This theorem can be stated as follows: "the truss can support the given loads if and only if, for all yield mechanisms, the power of dissipation is superior (i.e. $\geqslant$ 0) to the power of the given loads".

Let us consider, in the ith yield mechanism, (i = 1, ... μ), the power, called $\sigma_o b_i$, of the given loads, and the absolute value, called q_{ik}, of the elongation rate of the kth bar. Then, by the kinematic theorem we have

$$\sum_{k=1}^{\nu} q_{ik} s_k \geqslant b_i .$$

Calling Q the matrix (q_{ik}) and b the vector $\{b_i\}$, the stability condition can be written

$$(48) \qquad Qs \geqslant b .$$

I-11 **Statements of the Minimization Problems and of Their Dual Problems Mechanical Interpretation of Duality.**

The problem of the minimization of the mass is the following.

$$(49) \quad \begin{cases} \text{How to choose the cross section areas of the bars in such a way} \\ \text{that the truss supports the given loads and has a minimum mass.} \end{cases}$$

It should be remarked that minimizing the mass of the truss is equivalent to minimizing $\ell^T s$, where ℓ is the ν-vector of bar lengths.

I-11.1 Use of Stability Conditions (47)
Problem (49) can be stated

$$\begin{cases} \underline{\text{minimize}} \quad \ell^T s \\ \underline{\text{with}} \quad Ef = p \ , \ |f| \leqslant \sigma_o s \end{cases} \tag{50}$$

It is possible to reduce the number of unknowns. Indeed, if (s, f) is a solution of (50), then $|f| = \sigma_o s$; otherwise, it would be possible to reduce the cost without violating the constraints. Now, we decompose f into its positive and negative parts, f^+ and f^-. Then, each solution f of (50) is a solution of (51):

$$\begin{cases} \underline{\text{minimize}} \quad (\ell^T \ell^T) \begin{pmatrix} f^+ \\ f^- \end{pmatrix} \\ \underline{\text{on}} \ F_{1P} \subset R^{2\nu} : \ (E - E) \begin{pmatrix} f^+ \\ f^- \end{pmatrix} = p \ ; \ \begin{pmatrix} f^+ \\ f^- \end{pmatrix} \geqslant 0 \end{cases} \tag{51}$$

<u>Conversely</u>, in each solution of (51), either f_k^+ or f_k^- is equal to zero $(k = 1, \dots \nu)$. Otherwise, by reducing to zero the smallest of the two numbers f_k^+, f_k^- while keeping constant their difference, we would reduce the cost without violating the constraints. Then, every solution (f^+, f^-) of (51) gives a solution of (50) that is

$$f = f^+ - f^- \ , \quad s = \frac{1}{\sigma_o} (f^+ + f^-)$$

Finally, (51) is a statement of the minimization of the mass of the truss. It is a problem P_1 (cf. 4), with $n = 2\nu$, $c^T = (\ell^T \ell^T)$,

$$A = (E \ \ -E) \ , \ b = p \ , \ x = \begin{pmatrix} f^+ \\ f^- \end{pmatrix}$$

I-11.2 Use of Stability Condition (48)
Problem (49) can be stated

$$\begin{cases} \underline{\text{minimize}} \quad \ell^T s \\ \underline{\text{on}} \ F \subset R^\nu : \ Qs \geqslant b \ , \ s \geqslant 0 \end{cases} \tag{52}$$

It is a problem P (cf. 3).

I-11.3 Dual Problems. Mechanical Interpretation.   Using the general definitions of dual problems given in (I-8), we obtain the statements of the dual problems of (51) and (52) as follows:

$$(53) \begin{cases} \underline{\text{Dual of}} \ (51) - \underline{\text{maximize}} \ p^T y \\[2mm] \underline{\text{on}} \ \ F_{1D} \subset R^m : \begin{pmatrix} E^T \\ -E^T \end{pmatrix} y \leqslant \begin{pmatrix} \ell \\ \ell \end{pmatrix} \end{cases}$$

$$(54) \begin{cases} \underline{\text{Dual of}} \ (52) - \underline{\text{maximize}} \ b^T z \\[2mm] \underline{\text{on}} \ \ F_D \subset R^\mu : \ Q^T z \leqslant \ell \ , \ z \geqslant 0 \end{cases}$$

It is possible to give a mechanical interpretation of these dual problems. For example, let us examine problem (53). By comparison to (2), we set

$$y = \alpha \dot{u}$$

where α is a given strictly positive constant (time constant). Then, according to (40), dual problem (53) becomes

$$(55) \begin{cases} \underline{\text{maximize}} \ \ p^T \dot{u} \\[2mm] \underline{\text{with}} \ \ -\dfrac{\ell}{\alpha} \leqslant (\dot{\Delta \ell}) \leqslant \dfrac{\ell}{\alpha} \end{cases}$$

Thus, problem (53) has the following mechanical interpretation, which can be said to be the dual problem of (49):

$$(56) \begin{cases} \underline{\text{How to choose the velocity vector}} \ \dot{u} \ \underline{\text{of the joints such that the}} \\ \underline{\text{corresponding elongation rate vector}} \ (\dot{\Delta \ell} \ ) \ \underline{\text{of the bars verifies}} \\[2mm] \qquad -\dfrac{\ell}{\alpha} \leqslant \dot{\Delta \ell} \leqslant \dfrac{\ell}{\alpha} \qquad (\alpha > 0, \ \underline{\text{const.}}) \\[2mm] \underline{\text{and such that the power of the given loads is maximum.}} \end{cases}$$

I-12 Some Properties of Optimal Trusses

These properties are straightforward consequences of the general theorems given in subchapters A and B.

I-12.1 Basic Solutions of Primal Problem (49) (cf. I-3 and I-4)

> If problem (49) has a solution, then it has a basic solution (in
> every basic solution the $(\nu - m)$ non basic bars have a null cross   (57)
> section area). Every basic solution is statically determined.

I-12.2 Relations Between Solutions of Problems (49) and (56) (cf. (37) and (38))

> If one of problems (49) and (56) has a solution, so has
> the second, and the following equality holds   (58)
> $1/\rho$ (minimum of the mass) $= \alpha$ (maximum of the power of the
> given loads)
> (where ρ is the volumic density of mass)

I-12.3 Elongation Rate Vector of Non Null Bars Solutions of Primal Problem (49) when the Velocity Vector of the Joints is Solution of Dual Problem (56) (cf. (39))

> In each solution of dual problem (56), the elongation rate of
> each bar of any solution of (49) with a non null cross section   (59)
> area is equal to $\pm \ell_k /\alpha$, where ℓ_k is the initial length of this bar.

I-12.4 A Virtual Fail of Solutions of Primal Problem (49)

> If the strictly positive constant α is given and if problem (49)
> has a solution, then there exists a velocity vector of the joints
> such that the absolute value of the elongation rate of the kth (60)
> bar of the solution $(k = 1, ... \nu)$ is equal to ℓ_k /α if this bar has a
> non null section area and is inferior ($\leqslant$) to ℓ_k /α if this bar has
> a null cross section area.

This property is a consequence of Kuhn-Tucker theorem for problem P_1 which
can easily be deduced from (27).

REFERENCES

[1]   DANTZIG G.B. : *Linear Programming and Extensions*, New Jersey
           University Press, 1963.

[2]   GASS S.I.: *Linear Programming*, McGraw Hill, 2nd edition, 1964.

[3]   HADLEY G.: *Linear Programming*, Addison-Wesley, 1962.

[4]   MAIER G., SRIVIVASAN R., SAVE M.: *On Limit Design of Frames
           Using Linear Programming*, International Symposium on Computer-Aided
           Design, University of Warwick, England 10—14 July 1972.

[5]   SIMONNARD M.: *Programmation Linéaire*, Dunod, 1962.

CHAPTER II

NONLINEAR OPTIMIZATION

This chapter presents general methods of nonlinear optimization useful in structural optimization. It is successively dealing with main properties of convex sets and convex or concave functions, basic theorems of optimization, optimization problems of various kinds either without constraints or with equality constraints and inequality constraints, duality theory. In particular, Lagrange theorem, isoperimetric problems, Euler equations, Kuhn-Tucker theorems, saddlepoint theorems are treated. Several problems of mechanics illustrate these methods and show their importance.

A. CONVEXITY

II-1 Convex Sets
II-1.1

E_1 |
Definition of a Convex Set
A subset S of an affine space ϵ is convex if and only if, for all points $\bar{x}$, x of S, the segment $[\bar{x}, x]$ is in S.

It should be remarked that every point u of the segment $[\bar{x}, x]$ can be written

$$u = \bar{x} + \theta(x - \bar{x}) = (1 - \theta)\bar{x} + \theta x \ , \ \text{with} \ \ 0 \leqslant \theta \leqslant 1$$

Examples (The proof are easy and left to the reader).
i) The space ϵ is convex. By convention, the empty set is convex. The interior and the closure of a convex set in a normed space* are convex.

(*) More generally, this property is true in vectorial topological spaces. But most spaces of the mechanics are normed. So, we shall systematically use normed spaces.

ii) Let ℓ be a non null linear form on ϵ and let α be a number. Then the plane $\ell x = \alpha$ and the open (closed) half-spaces provided by this plane, i.e. $\{x \mid \ell x < \alpha\}$ and $\{x \mid \ell x > \alpha\}$ ($\ell x \leqslant \alpha$ and $\ell x \geqslant \alpha$) are convex.

iii) Every intersection (finite or infinite) of convex sets in ϵ is convex. For instance, the sets of R^n : $Ax = b$; $Ax > b$; $Ax \geqslant b$ (A, (m, n) matrix; b, m-vector) are convex.

II-1.2 Separating Planes and Supporting Planes

E_2

Definition of a Separating Plane

A plane $\ell x = \alpha$ in ϵ is said to separate (strictly separate) two subsets S_1 and S_2 of ϵ if and only if

x in S_1 implies $\ell x \leqslant \alpha$ − or $\ell x \geqslant \alpha$ ($\ell x < \alpha$ − or $\ell x > \alpha$)

x in S_2 implies $\ell x \geqslant \alpha$ − or $\ell x \leqslant \alpha$ ($\ell x > \alpha$ − or $\ell x < \alpha$)

Then, every one of the sets S_1, S_2 is contained in one of the closed half-spaces produced by the plane.

E_3

Definitions of a Supporting Plane, of a Supporting Point

A plane containing a point $\bar{x}$ of a subset S of ϵ is a supporting plane of S at $\bar{x}$, if and only if, all of S lies in one closed half-space produced by the plane.

A point $\bar{x}$ of S is a supporting point of S if and only if there exists a supporting plane of S at this point.

Given a supporting point $\bar{x}$, there may be more than a single supporting plane at $\bar{x}$.

There are fundamental theorems of separation of convex sets in ϵ. We are going to give some of these theorems when ϵ is R^n (without proofs).

E_4

Separation Theorem in R^n

Let S_1 and S_2 be two non empty disjoint convex sets in R^n.
There exists a plane which separates them.

E_5 | Strict Separation Theorem in R^n
Let S_1 and S_2 be two non empty convex sets in R^n, with S_1 compact and S_2 closed. Then S_1 and S_2 are disjoint if and only if there exists a plane which strictly separates them.

E_6 | Support Theorem in R^n
If $\bar{x}$ is a boundary point of a closed convex set S in R^n, then S has a supporting plane at $\bar{x}$.

The above theorems can be used to derive some theorems of the alternative, for instance Farkas theorem. We shall need support theorem in the necessary saddle point condition (II–16).

II-2 Convex and Concave Functions Without Differentiability
II-2.1

E_7 | Definitions
Let f be a numerical function on a convex set C in an affine space ϵ.
  i) The function f is convex (strictly convex) if and only if
      $f[(1 - \theta)\bar{x} + \theta x] \leqslant (1 - \theta)f(\bar{x}) + \theta f(x), 0 \leqslant \theta \leqslant 1$, for all
      points $\bar{x}, x$ of C
      $(f[(1 - \theta)\bar{x} + \theta x) < (1 - \theta)f(\bar{x}) + \theta f(x), 0 < \theta < 1$, for all
      points $\bar{x}, x$ with $\bar{x} \neq x$ of C)
  ii) The function f is concave (strictly concave) if and only if the
      function $(-f)$ is convex (strictly convex).

For instance, every linear or affine numerical function defined on a convex set in R^n is both convex and concave, but neither strictly convex nor strictly concave.

II-2.2 First Properties

E_8 | Linear Positive Combination of Convex Functions
Let f_1 and f_2 be two numerical functions on a convex set C in an affine space ϵ, and let λ_1 and λ_2 be two positive ($\geqslant 0$) numbers. Let us consider the function $f = \lambda_1 f_1 + \lambda_2 f_2$. Then

E_8

i) the function f is convex,

ii) if f_1 is strictly convex and if λ_1 is strictly positive, the function f is strictly convex.

<u>Proof of ii) (for instance)</u>

$$f\,[(1-\theta)\bar{x} + \theta x] = \Sigma\,\lambda_i\,f_i\,[\,(1-\theta)\bar{x} + \theta x\,]$$

$$< \Sigma\,\lambda_i\,[\,(1-\theta)f_i\,(\bar{x}) + \theta f_i\,(x)\,] = (1-\theta)f(\bar{x}) + \theta f(x)$$

<u>Definition of the Epigraph of f</u>

(1)

The epigraph of a given numerical function f on a set S in an affine space ϵ is the set

$$\{x,\zeta\}\subset S \times R \mid f(x) \leqslant \zeta$$

In other words, the epigraph of f is the set of the points of $S \times R$ which are above the graph of f.

<u>Convexity and Epigraph</u>

(2)

A numerical function f defined on a convex part C in an affine space ϵ is convex, if and only if, its epigraph is convex.

<u>Proof of the necessity</u>. Let (x^1, ζ^1) and (x^2, ζ^2) be two points of the epigraph of f. By convexity of f we have that for $0 \leqslant \theta \leqslant 1$

$$f[\,(1-\theta)x^1 + \theta x^2] \leqslant (1-\theta)f(x^1) + \theta f(x^2) \leqslant (1-\theta)\zeta^1 + \theta\zeta^2$$

Hence the point $(1-\theta)(x^1, \zeta^1) + \theta(x^2, \zeta^2)$ is in the epigraph of f. Then, this epigraph is convex.

<u>Proof of the sufficiency</u>. Let x^1, x^2 be two arbitrary points of C. Then $(x^1, f(x^1)), (x^2, f(x^2))$ are in the epigraph of f. By convexity of this epigraph we have

$$f\,[(1-\theta)x^1 + \theta x^2] \leqslant (1-\theta)f(x^1) + \theta f(x^2)$$

and hence f is convex.

Families of Convex Functions

(3) Let $\{f_i\}$ be a family (finite or infinite) of numerical convex functions on a convex set in ϵ. If these functions are bounded from above*, the least upper bound (that is the function f defined by $f(x) = \sup f_i(x)$) is convex.

Proof. This theorem is an immediate consequence of the theorem on epigraph (2) and of the theorem on intersection (II − 1.1 iii).

We end this section by giving a continuity theorem without proof.

Continuity of Convex Functions in R^n

(4) Every convex function on an open convex set in R^n is continuous.

II-3 Convex and Concave Functions with Differentiability

We shall establish general results for convex function. We would treat the concave functions case by changing the inequality signs $\geqslant$ and > 0 into $\leqslant$ and $<$.

II-3.1

Differentiable Functions

E9 Let f be a numerical differentiable function on an open convex set C in a normed space, and let $f'(\bar{x})$ be the differential (linear function) of f at $\bar{x}$. A necessary and sufficient condition for f be convex (strictly convex) is that

$$f(x) - f(\bar{x}) \geqslant f'(\bar{x})(x - \bar{x}) \quad \text{for all } \bar{x}, x \text{ in } C$$

$$(f(x) - f(\bar{x}) > f'(\bar{x})(x - \bar{x}) \quad \text{for all } \bar{x}, x \text{ with } \bar{x} \neq x \text{ in } C$$

***

(*) The theorem would be true without the assumption "bounded from above" if the functions f_i, f take their values into R.

Geometrical Interpretation

The linearization $f(\bar{x})+f'(\bar{x})(x - \bar{x})$ always underestimates $f(x)$ for every x (strictly underestimates for $x \neq \bar{x}$). (see figure).

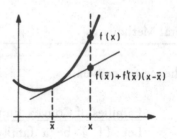

Proof (for instance, for non strict convexity).

Necessity. According to convexity we have, for $0 < \theta \leqslant 1$,

$$f(x) - f(\bar{x}) \geqslant \frac{1}{\theta} (f[\bar{x} + \theta(x - \bar{x})] - f(\bar{x}))$$

Let us consider the numerator of the right-hand side as a function of θ and apply Taylor theorem and the linearity of the differential. The right-hand side becomes

$$f'(\bar{x})(x - \bar{x}) + \epsilon(\bar{x}, \theta(x - \bar{x}))\| x - \bar{x} \|$$

where the number $\epsilon(\bar{x}, \theta(x - \bar{x}))$ tends to zero when θ tends to zero. The inequality is still true at the limit. Then we obtain

$$f(x) - f(\bar{x}) \geqslant f'(\bar{x})(x - \bar{x})$$

Sufficiency. Let $\bar{x}$ and x be two arbitrary points of C. The point

$$u = \bar{x} + \theta(x - \bar{x}) \quad , \quad 0 \leqslant \theta \leqslant 1$$

is in C, and the inequality assumption applied successfully to $\bar{x}$, u and to x, u gives

$$f(\bar{x}) - f(u) \geqslant - \theta f'(u)(x - \bar{x})$$

$$f(x) - f(u) \geqslant (1 - \theta) f'(u)(x - \bar{x})$$

Multiplying the first inequality by $(1 - \theta)$, the second one by θ, and adding, give

$$(1 - \theta)f(\bar{x}) + \theta f(x) \geqslant f[(1 - \theta)\bar{x} + \theta x]$$

Then, f is convex.

II-3.2

> **Twice Differentiable Functions**
> Let f be a numerical twice differentiable function on an open
> convex set C in a normed space, and let f''(x) be the second
> differential (bilinear function) of f at the point x of C.
> E_{10} i) If f is convex, then f''(x) is positive semidefinite and
>   conversely if f''(x) is positive semidefinite for each x of C,
>   then f is convex.
>   ii) If f is strictly convex, then f''(x) is not necessarily positive
>   definite, but if f''(x) is positive definite for each x of C, then f
>   is strictly convex.

Proofs

i) <u>Necessity</u>. Because C is open, there exists $\delta > 0$ such that, for each h
having the norm δ the point $x + th$, with $0 < t < 1$, is in C. By
theorem E_9 it follows that

$$f(x + th) - f(x) - tf'(x)h \geqslant 0$$

But since f is twice differentiable at x, the left-hand side is

$$\frac{t^2}{2} f''(x)(h,h) + t^2 \eta(x,th)\delta^2$$

Hence we have

$$f''(x)(h,h) + 2\eta(x,th)\,\delta^2 \geqslant 0$$

Taking the limit when t tends to zero, we obtain

$$f''(x)(h,h) \geqslant 0$$

This inequality holds for $\| h \| = \delta > 0$, and therefore for all vectors h:
the second differential f''(x) is positive semidefinite.

i) <u>Sufficiency.</u> By Taylor theorem we have for each $\bar{x}$ and x of C

$$f(x) - f(\bar{x}) - f'(\bar{x})(x - \bar{x}) = \frac{1}{2} f''(\bar{x} + \lambda(x - \bar{x}))(x - \bar{x}, x - \bar{x})$$

for some λ, $0 < \lambda < 1$. The right-hand side is positive, and therefore, by E_9 the function f is convex.

ii) <u>Necessity.</u> Counterexample: $\epsilon = C = R$, $f(x) = x^4$, $f''(x) = 12x^2$. The function f is strictly convex, but $f''(x)$ is not positive definite.

ii) Sufficiency. The proof is similar to i) sufficiency.

II-3.3

<u>Example.</u> $\epsilon = C = R^n$, $f(x) = 1/2\, x^T A x$, where $\bar{A}$ is $a(n,n)$—matrix. The second differential is

$$f''(x) = A$$

Then, if A is a positive semidefinite matrix, f is convex; if A is a positive definite matrix, f is strictly convex.

B. DEFINITIONS AND BASIC PROPERTIES OF OPTIMIZATION WITHOUT DIFFERENTIABILITY

II-4 Definitions About Minima and Maxima

E_{11}

<u>Definition of a Minimum</u>
A numerical function f defined on an arbitrary set S has a minimum (resp. a strict minimum) if and only if there exists an element $\bar{x}$ of S such that

$$f(\bar{x}) \leqslant f(x) \qquad \text{for all x of S}$$

$$(\text{resp. } f(\bar{x}) < f(x) \quad \text{for all } x \neq \bar{x} \text{ of S})$$

E_{12}

<u>Definition of an infimum</u>
i) A numerical function f defined on an arbitrary set S has an infimum α if and only if we have

E_{12}

 . $\alpha \leqslant f(x)$ for all x of S,
 . for every $\epsilon > 0$ there exists an x of S such that $f(x) < \alpha + \epsilon$
   ii) A numerical function f defined on an arbitrary set S has an
       infimum equals to $-\infty$, if and only if, for every number a
       there exist points of S such that $f(x) < a$.

The definitions of <u>maximum</u> and <u>supremum</u> are obtained from the corresponding definitions of minimum and infimum by reversing the inequality signs.

<u>Remarks</u>
 — The minimum of a numerical function, if it exists, is an attained infimum.
 — The minimum of a numerical function, if it exists, is finite.

E_{13}

Definition of a Local Minimum
A numerical function f defined on a set S in a normed space has a
local minimum at $\bar{x}$, if and only if there exists $\delta > 0$ such that f
has a minimum over the intersection of S and of the ball
$\| x - \bar{x} \| < \delta$.

Consequently, if f has a minimum at $\bar{x}$, it has a local minimum at this point. Of
course, the converse is not true.

II-5 Existence Theorems

E_{14}

Definition of a Lower Semicontinuous Function
A numerical function f defined on a set S in a normed space is said
to be lower semicontinuous (l.s.c.) at $\bar{x}$, if and only if, for each
$\epsilon > 0$, there exists a $\delta > 0$ such that

$$f(\bar{x}) - \epsilon < f(x)$$

for all x in the ball $\| x - \bar{x} \| < \delta$.
The function f is said to be lower semicontinuous on S, if and only
if it is lower semicontinuous at every point of S.

The function f is said to be upper semicontinuous (u.s.c.) on S if and only if
the function $(-f)$ is l.s.c. on S.

It is proved that f is continuous if and only if it is both l.s.c. and u.s.c.
It is also proved that f is l.s.c. if and only if its epigraph is closed.

E_{15}
Existence of Minimum
Let f be a continuous numerical function on a non-empty closed set
S in an Hilbert space*. If S is bounded, or if $f(x)$ tends to $+ \infty$ when
$\| x \|$ tends to $+ \infty$, then the function f has a minimum.

E_{16}
Existence of Infimum
Every numerical function defined on a non empty set in an
arbitrary space has an infimum (and a supremum) finite or
infinite.

II-6 Convexity and Concavity Cases
II-6.1

E_{17}
Minimization of a Convex Function
Let f be a convex numerical function on a convex set C in a
normed space. Then
i) if f has a local minimum at $\bar{x}$, it has a minimum at
   this point.
ii) the set of points where f is minimum is convex.
iii) if f is strictly convex, there exists at most one point where it is
   minimum.

Proof of i). According to the definition of a local minimum, there exists
$\delta > 0$ such that f has at $\bar{x}$ a minimum on the intersection C' of C and of the ball
$\| x - \bar{x} \| < \delta$.
Let x be an arbitrary point of C, different from $\bar{x}$. The point

$$u = \bar{x} + \theta (x - \bar{x})$$

(*) An Hilbert space is a vectorial space provided with an inner product such that, for the topology associated
with the corresponding norm, this space is complete (R^n is an Hilbert space).

with θ in the open interval

$$0 < \theta < \min (1, \delta / \| x - \bar{x} \|) \qquad (5)$$

is in C'. Therefore we have

$$f(\bar{x}) \leqslant f(u) \qquad (6)$$

On the other hand, the convexity of f implies

$$f(u) \leqslant (1 - \theta) f(\bar{x}) + \theta f(x) \qquad (7)$$

Finally, from (6), (7) and (5), we deduce

$$f(\bar{x}) \leqslant f(x)$$

This inequality is true for each point x of C. Then f has a minimum at $\bar{x}$.

Proof of ii). Let $\bar{x}$ and $\bar{\bar{x}}$ be two points where f is minimum. Because convexity, at every point $u = (1 - \theta) \bar{x} + \theta \bar{\bar{x}}$, $0 \leqslant \theta \leqslant 1$, we have

$$f(u) \leqslant (1 - \theta) f(\bar{x}) + \theta f(\bar{\bar{x}})$$

Since f is minimum at $\bar{x}$ and at $\bar{\bar{x}}$, we have

$$f(\bar{x}) = f(\bar{\bar{x}})$$

and the previous inequality implies

$$f(u) = f(\bar{x})$$

Hence, f is minimum at every point of the segment $[\bar{x}, \bar{\bar{x}}]$: the set of the points where f is minimum is convex.

Proof of iii). Let $\bar{x}$ and $\bar{\bar{x}}$ ($\bar{x} \neq \bar{\bar{x}}$) be two points where f is minimum. Then, according to ii), on all points of $]\bar{x}, \bar{\bar{x}}[$ the function f takes the same value:$f(\bar{x})$; this contradicts the assumption that f is strictly convex.

II-6.2

E_{17}' | **Minimization of a Concave Function**
Let f be a nonconstant concave numerical function on a convex part C of a normed space. If f has a minimum at $\bar{x}$, then $\bar{x}$ is a boundary point.;

<u>Proof</u>. We are going to show that, at every interior point x, we have $f(x) > f(\bar{x})$.

Because f is non constant and has a minimum at $\bar{x}$, there exists a point y of C such that

$$(7') \quad f(y) > f(\bar{x})$$

Let x be an arbitrary interior point. There exist a point z of C and a number θ, $0 \leqslant \theta \leqslant 1$, such that

$$x = y + \theta(z - y)$$

Then

$$f(x) \geqslant (1 - \theta)f(y) + \theta f(z) \quad , \quad \text{(concavity of f)}$$

$$> (1 - \theta)f(\bar{x}) + \theta f(\bar{x}) \quad , \quad ((7') \text{ with } 1 - \theta > 0, f(y) > f(\bar{x}))$$

$$= f(\bar{x})$$

Hence, the theorem is proved.

C. UNCONSTRAINED OPTIMIZATION PROBLEMS (WITH DIFFERENTIABILITY)

All this subchapter is dealing with the following problem.

E_{18} | **Problem.** Minimize or maximize a differentiable (possibly twice differentiable) numerical function f on an <u>open</u> set S in a normed space ϵ.

II-7 Necessary Conditions

II-7.1

E_{19} | **Differentiable Functions**
Suppose that the function $f(E_{18})$ is differentiable and let $\bar{x}$ be a point of S. If f has a local minimum or a local maximum at $\bar{x}$, then $f'(\bar{x}) = 0$.

Remarks

i) The condition $f'(\bar{x}) = 0$ means that the number $f'(\bar{x})h$ is zero for all vectors h. If ϵ is R^n, it can be written $\nabla f(\bar{x}) = 0$.

ii) The theorem would not be true if S would not be an open set. Example: in R, $f(x) = x$ on $[0,1]$, $\bar{x} = 0$.

iii) The condition $f'(\bar{x}) = 0$ is not sufficient for f be minimum or maximum at $\bar{x}$. Example: in R^2 , $f(x) = x_1^2 - x_2^2$, $\bar{x} = 0$.

Proof. Since S is open and f has a local minimum or a local maximum at $\bar{x}$, there exists an open ball $(\bar{x}, \delta)$ contained in S on which f has a minimum or a maximum at $\bar{x}$.

Given a non null vector h, every point of the straight line $(\bar{x}, h)$ which is in the previous ball can be written

$$x = \bar{x} + th \quad \text{with} \quad |t| < \delta/\|h\|$$

The numerical function

$$g(t) = f(\bar{x} + th) \tag{8}$$

has a minimum or a maximum at $t = 0$, it is differentiable at this point. Therefore its derivative is null

$$f'(\bar{x})h = 0$$

Hence, the theorem (E_{19}) is proved.

II-7.2

E_{20} | Twice Differentiable Functions
Suppose that f is twice differentiable and let $\bar{x}$ be a point of S. If f
has a local minimum (local maximum) at $\bar{x}$, then $f''(\bar{x})$ is
semidefinite positive (semidefinite negative).

Proof. The numerical function g (8) has a minimum (a maximum) at $\bar{x}$, it is twice
differentiable at this point. Then its second derivative is positive (negative):

$$f''(\bar{x})(h,h) \geqslant 0 \qquad (f''(\bar{x})(h,h) \leqslant 0)$$

that is to say $f''(\bar{x})$ is semidefinite positive (negative).

Remarks

i)   If ϵ is R^n the condition $f''(\bar{x})$ semidefinite positive means $h^T H(\bar{x})h \geqslant 0$
     for all vectors h ($H(\bar{x})$ is the hessian of f at $\bar{x}$).

ii) We would presume that, if the minimum is strict, then $f''(\bar{x})$ is definite
     positive. This is not so. Example: in R^2, $f(x) = x_1^2 + x_2^4$, $\bar{x} = 0$.

II-8 Sufficient Condition

E_{21} | Twice Differentiable Functions
Suppose that f is twice differentiable, and let $\bar{x}$ be a point of S. If
$f'(\bar{x}) = 0$ and if $f''(\bar{x})$ is coercive, then f has a strict local minimum
at $\bar{x}$.

E_{22} | Definition of Coercivity
The second differential $f''(\bar{x})$ is coercive if and only if there exists
$\alpha > 0$ such that

$$f''(\bar{x})(h,h) \geqslant \alpha \| h \|^2 \quad \text{for all vectors h}$$

If $f''(\bar{x})$ is coercive, then it is definite positive. It is proved that if the dimension
of ϵ is finite the converse is true, and if the dimension of ϵ is not finite, the
converse is not true.

Proof of the theorem E_{21} — Because S is open, there exists an open ball
centred at $\bar{x}$, having β for radius, contained in S. Then Taylor theorem gives

$$f(\bar{x} + h) - f(\bar{x}) = \frac{1}{2} f''(\bar{x})(h,h) + \epsilon(h)\| h \|^2$$

with limit $\epsilon(h) = 0$ when h tends to zero.

By coercivity of $f''(\bar{x})$ it follows

$$f(\bar{x} + h) - f(\bar{x}) \geqslant \frac{1}{2}(\alpha + 2\epsilon(h))\| h \|^2$$

Now, there exists δ, $0 < \delta < \beta$ such that $\| h \| < \delta$ implies $\alpha + 2\epsilon(h) > 0$.

Thus the function f does have a strict local minimum at x.

II-9 Convexity Case

E_{23} | Let f be a numerical differentiable convex (strictly convex) function on an open convex set C in a normed space, and let $\bar{x}$ be a point of C. The function f has a minimum (a strict minimum) at $\bar{x}$ if and only if $f'(\bar{x}) = 0$.

Proof. If f has a minimum at $\bar{x}$, it has a local minimum at this point, and $f'(\bar{x}) = 0$ by theorem E_{19}

Conversely, if $f'(\bar{x}) = 0$, by theorem E_9 we have $f(x) \geqslant f(\bar{x})$ for all points x of C.

Thus the function f is minimum at $\bar{x}$.

If f is strictly convex, this minimum is strict (E_{17}).

II-10 Euler Equations

Let us denote E the vectorial space of the function φ defined on the segment $I = [a, b]$ in R, $a < b$, taking their values into R^n, continuous, piecewise differentiable (the derivative vector $\dot{\varphi}(t)$ is an element of R^n), provided with the norm

$$\|| \varphi \|| = \sup_{t \in I} (\|\varphi(t)\| + \|\dot{\varphi}(t)\|)$$

It should be remarked that the dimension of E is infinite.

In several problems of mechanics we have to introduce the subset S of E made by the functions φ of E such that the point $\{\varphi(t), \dot{\varphi}(t)\}$ is in a given domain (open connex subset) D in R^{2n}.

From every number t of I and from every function φ of E, we form the number $L[t, \varphi(t), \dot{\varphi}(t)]$ where L is a given function, differentiable with respect to the first variable, twice differentiable with respect to the second variable (φ) and the third variable ($\dot{\varphi}$).

Finally, we define on S a numerical function f by

(9)
$$f(\varphi) = \int_a^b L[t, \varphi(t), \dot{\varphi}(t)] \, dt$$

The problem is to minimize or to maximize f on S.

It would be pleasant to apply the theorem E_{19}. Fortunately the subset S is open in E and the function f is differentiable on S, just as you are going to prove.

i) <u>S is an open subset of</u> E. Let φ be a given function of S. When t lies in I, the point $\{\varphi(t), \varphi(t)\}$ describes a closed subset $\sqcap$ in D. Let γ be the "distance" (associated with the norm) from $\sqcap$ to the complementary subset of D in R^{2n}. Because this complementary subset and $\sqcap$ are closed and disjoint, the distance γ is strictly positive. Let ψ an arbitrary function of E such that

$$\|\!|\!| \psi - \varphi |\!|\!|\! < \frac{\gamma}{2}$$

Then the distance from $\{\psi(t), \dot{\psi}(t)\}$ to $\{\varphi(t), \dot{\varphi}(t)\}$ in R^{2n} is strictly smaller than $\gamma/2$. Therefore the point $\{\psi(t), \dot{\psi}(t)\}$ is in D. Hence ψ is in S.

ii) <u>f is differentiable on</u> S.
The differential $f'(\varphi)$, if it exists, is a continuous linear function such that, for each $\epsilon > 0$, there exists $\delta > 0$ such that

$$\|\!|\!| h |\!|\!|\! < \delta \quad \text{implies} \quad |f(\varphi + h) - f(\varphi) - f'(\varphi)h \leqslant \epsilon \|\!|\!| h |\!|\!|$$

We apply Taylor theorem to the function L: given $\eta > 0$, there exists $\delta_0 > 0$ such that $|\!|\!| h |\!|\!| < \delta_0$ implies

$$\left| L\left[t,(\varphi + h)(t), (\dot{\varphi} + \dot{h})(t)\right] - L\left[t,\varphi(t),\dot{\varphi}(t)\right] \right.$$

$$\left. - L'_{\varphi}\left[t,\varphi(t),\dot{\varphi}(t)\right]h(t) - L'_{\dot{\varphi}}\left[t,\varphi(t),\dot{\varphi}(t)\right]\dot{h}(t)\right| \leqslant \eta \parallel\!\mid h \mid\!\parallel$$

and, with greater reason,

$$\left| f(\varphi + h) - f(\varphi) - \mathcal{L}(\varphi)h \right| \leqslant \eta(b - a) \parallel\!\mid h \mid\!\parallel \ , \ \text{with}$$

$$\mathcal{L}(\varphi)h = \int_a^b \left(L'_{\varphi}\left[t,\varphi(t), \dot{\varphi}(t)\right]h(t) - L'_{\dot{\varphi}}\left[t,\varphi, \dot{\varphi}(t)\right]\dot{h}(t)\right)dt \qquad (10)$$

It remains to be shown that $\mathcal{L}(\varphi)$ is continuous, i.e. it is bounded, that is to say

$$\left| \mathcal{L}(\varphi)h \right| < M \parallel\!\mid h \mid\!\parallel \ , \quad \text{for all } h \text{ in } E,$$

where M is a fixed number.

Now, according to hypothesis, there exists K such that

$$\left| \mathcal{L}(\varphi)h \right| \leqslant K \int_a^b \left(\parallel h(t) \parallel + \parallel \dot{h}(t) \parallel \right)dt \ \leqslant K \int_a^b \parallel\!\mid h \mid\!\parallel dt = K(b-a)\parallel\!\mid h \mid\!\parallel$$

Hence, $\mathcal{L}(\varphi)$ is bounded, and continuous.

Finally, the function f is differentiable on S and the differential $f'(\varphi)$ is $\mathcal{L}(\varphi)$ (10); it is obtained from (9) by differentiation under the integral sign.

Integrating by parts permits to write

$$f'(\varphi)h = L'_{\dot{\varphi}}\left[t,\varphi(t),\dot{\varphi}(t)\right]h(t) \Bigg|_{t=a}^{t=b} + \int_a^b \left(L'_{\varphi}\left[t,\varphi(t),\dot{\varphi}(t)\right]\right.$$

$$\left. - \frac{d}{dt} L'_{\dot{\varphi}}\left[t,\varphi(t),\dot{\varphi}(t)\right]\right)h(t)dt \qquad (11)$$

that is to say, in abbreviated form,

$$\delta f = \frac{\partial L}{\partial \dot{\varphi}} \delta\varphi \Bigg|_{t=a}^{t=b} + \int_a^b \left(\frac{\partial L}{\partial \varphi} - \frac{d}{dt} \frac{\partial L}{\partial \dot{\varphi}} \right) \delta\varphi \ dt \qquad (12)$$

δf is said to be the first variation of f.

iii) <u>Necessary optimality condition</u>. Now, let us suppose that f has a local minimum or a local maximum for the function $\bar{\varphi}$. Then f'($\bar{\varphi}$) = 0 and the number f'($\bar{\varphi}$)h (11) is zero for all functions h of E. With greater reason, this number is zero for all functions h of E verifying

$$h(a) = h(b) = 0$$

For these functions we have from (11)

(13)
$$\int_a^b \left(\frac{\partial L}{\partial \varphi} - \frac{d}{dt} \frac{\partial L}{\partial \dot{\varphi}} \right) h(t) dt = 0$$

The operator

$$\frac{\partial L}{\partial \varphi} - \frac{d}{dt} \frac{\partial L}{\partial \dot{\varphi}}$$

has n components which will be noted $\mathscr{P}_j (t)$. Hence, the identity (13) takes the form

(14)
$$\Sigma \int_a^b \mathscr{P}_j(t) h_j (t) = 0 \qquad (j = 1, ... \ n)$$

It is deduced from (14) that all the $\mathscr{P}_j$ are null. Indeed, suppose for instance $\mathscr{P}_k (t) > 0$ for $t = t_0$. According to continuity of $\mathscr{P}_k$ we have $\mathscr{P}_k (t) > 0$ on a closed interval $[t_1 \ t_2]$, $t_1 \neq t_2$, containing t . Let us take for h a continuous function such that

$$h_j (t) = 0 \qquad for \qquad j \neq k$$

$$h_k(t) \begin{cases} > 0 \text{ when } t \text{ is in an open interval contained in } [t_1 \ t_2] \text{ and which does} \\ \qquad \text{not contain neither a nor b} \\ = 0 \ \text{ anywhere else.} \end{cases}$$

Then the left-hand side of (14) is > 0; this contradicts (14).

(15) Finally, if the function f(9) has on S a local minimum or a local maximum for $\varphi = \bar{\varphi}$, then $\bar{\varphi}$ verifies the equations (Euler equations)

$$\left| \quad \frac{\partial L}{\partial \varphi_j} - \frac{d}{dt} \frac{\partial L}{\partial \dot{\varphi}_j} = 0 \qquad (j = 1, \dots n) \qquad (15)\right.$$

Remarks.

i) Of course, Euler equations are not sufficient to imply optimality.

ii) If the function L does not depend on $\dot{\varphi}$, it is not necessary to suppose that the functions φ of E and continuous and piecewise differentiable; the piecewise continuity is sufficient.

iii) When the integrand of (9) is

$$L\left[t, \varphi(t), \dot{\varphi}(t), \ddot{\varphi}(t)\right]$$

the Euler equations are

$$\frac{\partial L}{\partial \varphi} - \frac{d}{dt} \frac{\partial L}{\partial \dot{\varphi}} + \frac{d^2}{dt^2} \frac{\partial L}{\partial \ddot{\varphi}} = 0 \qquad (16)$$

D. OPTIMIZATION WITH EQUALITY CONSTRAINTS

E24
| Notations
| S: open set in a normed space ϵ
| f, g_i (i = 1, ...m): numerical differentiable functions on S
| g: column matrix $\{g_i\}$
| F: set g(x) = 0 (F is in S)

II-11 Necessary Conditions

E25
| Lagrange Multipliers Theorem
| If f has on $F(E_{24})$ a local minimum or a local maximum at $\bar{x}$ and
| if the differentials $g'_i (\bar{x})$ are linearly independent, then there exist
| m numbers y_i (called Lagrange multipliers) such that
| $$f'(\bar{x}) = \sum_{i=1}^{m} y_i \, g'_i(\bar{x})$$

Proof

i) The dimension of ϵ is finite. Let n be this dimension. Without loss of generality, it can be supposed that the square matrix $\partial g_i / \partial x_j$ ($\bar{x}$), i and j = 1, ...m, is regular.

We use the following notations

$$u = (x_1, \ldots x_m)^T \quad , \quad \bar{u} = (\bar{x}_1, \ldots \bar{x}_m)^T$$

$$v = (x_{m+1}, \ldots x_n)^T \quad , \quad \bar{v} = (\bar{x}_{m+1}, \ldots \bar{x}_n)^T$$

Because f has on F a local minimum or a local maximum at $\bar{x}$, and according to the implicit function theorem, there exist on F a neighbourhood F' of $\bar{x}$, an open set Σ in R^{n-m} and a differentiable function φ : $u = \varphi(v)$, such that f has a minimum on F' at $\bar{x}$ and the point $x = (\varphi(v), v)^T$ lies on F' when v describes Σ.

Then, on the open set Σ the function $v \to f[\varphi(v), v]$ has a minimum or a maximum at $\bar{v}$. Therefore, by theorem E_{19}, we have

(17) $$f'_u(\bar{x})\varphi'(\bar{v}) + f'_v(\bar{x}) = 0$$

But the equality

$$g[\varphi(v), v] = 0$$

holds for all v in Σ, and therefore

(18) $$g'_u(\bar{x})\varphi'(\bar{v}) + g'_v(\bar{x}) = 0$$

Since $g'_u(\bar{x})$ is regular, we can calculate $\varphi'(\bar{v})$ from (18) and thus, by putting

$$y^T = f'_u(\bar{x})[g'_u(\bar{x})]^{-1} \quad , \quad (y \text{ is an element of } R^m)$$

we obtain

$$\begin{cases} f'_u(\bar{x}) = y^T g'_u(\bar{x}) \\ f'_v(\bar{x}) = y^T g'_v(\bar{x}) \end{cases}$$

that is the relation of the theorem E_{25} i.e.

$$f'(\bar{x}) = y^T g'(\bar{x})$$

ii) The dimension of ϵ is infinite. Since the differentials $g_i'(\bar{x})$ are independent, there exist vectors h^j such that

$$g_i'(\bar{x})h^j = \delta_{ij} \quad \text{(Kronecker symbol)} \quad (j = 1, ...m)$$

It is easy to be shown that these vectors h^j are independent.
Now, let us take

$$z_i = f'(\bar{x})h^i \quad , \quad (i = 1, ...m) \tag{19}$$

and

$$\mathcal{A} = f'(\bar{x}) - \sum_{i=1}^{m} z_i g_i'(\bar{x}) \tag{20}$$

By definition of z_i (19) we have

$$\mathcal{A}h^j = 0 \quad (j = 1, ...m)$$

Now, we are going to show that

$$\mathcal{A}h = 0 \quad \text{for each vector h of } \epsilon \tag{21}$$

Of course, if h is in the vectorial subspace of ϵ spanned by the vectors h^j, the relation (21) holds.

Otherwise, the space ϵ' spanned by the point $\bar{x}$ and the vectors $h^1...h^m,h$ has the dimension $(m+1)$. On ϵ' the function f has a local minimum or a local maximum at $\bar{x}$. Therefore, by theorem E_{25} previously proved in space of finite dimension, there exist m numbers y_i such that

$$\mathcal{B} = f'(\bar{x}) - \sum_{i=1}^{m} y_i g_i'(\bar{x})$$

is null on $h^1,...h^m,h$. Writing this condition for h^j we obtain

$$(f'(\bar{x}) - \sum_i y_i g_i'(\bar{x}))\, h^j = f'(\bar{x})h^j - y_j = 0$$

Comparing to (19) we deduce $z_i = y_i$ and, consequently, $\mathcal{A} = \mathcal{B}$.
Thus, the relation (21) is verified for every vector h.
The theorem is proved.

II-12 Isoperimetric Problems

We return to the assumptions made in (II-10) on the sets E, S and on the function L. Now, we introduce equality constraints as follows. Given m numerical functions M_i having the same properties than the function L, we define F as the intersection of S and of $g_i(\varphi) = C_i$, $i = 1,...m$, $(C_i, \text{const.})$ with

$$g_i(\varphi) = \int_a^b M_i[t, \varphi(t), \dot\varphi(t)]dt$$

We consider the problem: minimize or maximize the numerical function f(9) on F. The theorem E_{25} gives a necessary condition: suppose that $\bar\varphi$ is a solution and that the m differentials $g'_i(\bar\varphi)$ are linearly independent. Then, there exist m numbers y_i such that $\bar\varphi$ verifies the following equations

$$(22) \qquad \frac{\partial L}{\partial \varphi_j} - \frac{d}{dt}\frac{\partial L}{\partial \dot\varphi_j} = \sum_{i=1}^m y_i \left(\frac{\partial M_i}{\partial \varphi_j} - \frac{d}{dt}\frac{\partial M_i}{\partial \dot\varphi_j} \right), \quad j = 1,... n$$

It should be noted that there are (n + m) unknown quantities (n components ρ_j, m multipliers) and (n + m) equations (n Euler equations (22), m constraint equations $g_i(\bar\varphi) = C_i$).

Problems of previous types are called isoperimetric problems.

Euler equations with multipliers (22) are very useful in numerous problems on structural optimization (see subchapter F).

E. OPTIMIZATION WITH INEQUALITY CONSTRAINTS. DUALITY

E_{26}

Notations

· f, g_i (i = 1,...m), h_k (k = 1,...p): numerical functions defined on a given subset S of R^n.
· g: column-matrix $\{g_i\}$; h: column-matrix $\{h_k\}$.
· F: subset of S defined by $g(x) \geqslant 0$.
· F': subset of S defined by $g(x) \geqslant 0$ and $h(x) = 0$.

E_{27}

Problem P
Minimize f on F

E_{28} | $\underline{\text{Problem P'}}$
Minimize f on F'

The subsets F and F' of S are called the feasible regions (of the problems P and P' respectively). The points x of a feasible region are said to be feasible points. As in chapter I, we can transform any problem P' into a problem P by replacing the equality h(x) = 0 by the two inequalities h(x) $\geqslant$ 0, − h(x) $\geqslant$ 0, and any problem P into a problem P' by creating slack variables. Therefore we shall mainly study one of these problems, for example the one which has the simplest form, that is to say the problem P.

We shall state necessary conditions and sufficient conditions of optimality. In paragraphs (II-13) and (II-14) we shall suppose that the functions f,g,h have some properties of differentiability. These assumptions will be deleted in paragraphs (II-15) and (II-16). Paragraph (II-17) will be devoted to properties dealing with duality.

II-13 Necessary condition. Kuhn-Tucker Theorem

The Kuhn-Tucker theorem is similar to the La range multipliers theorem (here, the multipliers are positive). As for this theorem, some quite general properties, called qualification properties, must be imposed on the constraints. Numerous constraint qualifications have been stated. Here, we shall study those given by Kuhn and Tucker themselves.

To be able to express them conveniently, we introduce the two following concepts.

II-13.1 Locally constrained vectors. Attainable vectors

A vector u of R^n is locally constrained at the point $\bar{x}$ in F, if and only if, g is differentiable at $\bar{x}$ and $u^T \nabla g_i(\bar{x})$ is positive ($\geqslant 0$) for all i such that $g_i(\bar{x}) = 0$. The set of numbers i verifying $g_i(\bar{x}) = 0$ will be noted $N(\bar{x})$ (see (23), chapter I).

A vector u of R^n is attainable at the point $\bar{x}$ in F, if and only if, there exists an n-dimensional vector function ψ, t $\rightarrow \psi(t)$, defined on $[0, t_1]$, $t_1 > 0$. such that $\psi(0) = \bar{x}$, $g(\psi(t) \geqslant 0$, ψ is differentiable at t = 0 and $d\psi/dt(0) = u$. (This means that there exists a curve starting at $\bar{x}$ in the direction of u and remaining in the feasible region).

The relations between these two concepts are shown by the two following properties.

i) Every vector u which is attainable at $\bar{x}$ is also locally contained at $\bar{x}$. Indeed, if $\psi(t)$ is a vector function which makes u attainable at $\bar{x}$, we have

$$\frac{dg_i(\psi)}{dt}(0) = \nabla g_i(\bar{x}) \cdot \frac{d\psi}{dt}(0)$$

For each i of $N(\bar{x})$ the left-hand side is positive ($\geqslant 0$), because $g_i(\psi(t))$ is null for $t = 0$ and positive for $0 < t \leqslant t_1$. Therefore, the right-hand side, which can be written $u^T \nabla g_i(\bar{x})$, is also positive.

ii) But not every vector which is locally constrained at $\bar{x}$ is necessarily attainable there.

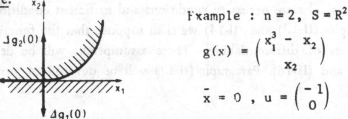

Example : $n = 2$, $S = R^2$

$$g(x) = \begin{pmatrix} x_1^3 - x_2 \\ x_2 \end{pmatrix}$$

$$\bar{x} = 0 \ , \ u = \begin{pmatrix} -1 \\ 0 \end{pmatrix}$$

The vector u is locally constrained at 0, but no curve can start from this point in the direction of u without leaving F immedmtcly.

II-13.2 Kuhn-Tucker Constraint Qualification

i) If every vector locally constrained at $\bar{x}$ is also attainable at this point, the constraint is said to satisfy the Kuhn-Tucker qualification.

We repeat this definition in such a way that we can refer to it.

E29

> #### Kuhn-Tucker Constraint Qualification
> Supposing that S is open, the function g is said to satisfy the Kuhn-Tucker qualification for problem P at a point $\bar{x}$ of F, if and only if, g is differentiable at $\bar{x}$ and, for every vector u of R^n satisfying the inequality $u^T \nabla g_i(x) \geqslant 0$ for all i such that $g_i(\bar{x}) = 0$, there exists an n-dimensional vector function $t \rightarrow \psi(t)$ defined on $[0, t_1]$, $t_1 > 0$, such that : $\psi(0) = \bar{x}$, $g(\psi(t)) \geqslant 0$, ψ is differentiable at $t = 0$ and $d\psi/dt(0) = u$.

If all the constraints are linear, it is not necessary to make sure that they satisfy the Kuhn-Tucker qualification. Indeed, we are going to demonstrate the following property.

E_{30} | If the constraints are linear, then the Kuhn-Tucker qualification holds at every point of the feasible region.

Proof. Let $Ax - b \geqslant 0$ be the constraints. If u is locally constrained at $\bar{x}$, that is to say if $\sum_j a_{ij} u_j \geqslant 0$ for all i belonging to $N(\bar{x})$, let us take

$$\psi(t) = \bar{x} + tu$$

First, $\psi(0) = \bar{x}$ and $d\psi/dt$ (0) = u. Then, let us consider the expression

$$\sum_j a_{\alpha j} \psi_\alpha(t) - b_\alpha = (\sum_j a_{\alpha j} \bar{x}_j - b_\alpha) + t \sum_j a_{\alpha j} \mu_j$$

If α is in $N(\bar{x})$, this expression is positive for all $t \geqslant 0$. If α is not in $N(\bar{x})$, it is positive for $0 \leqslant t \leqslant t_1$ where t_1 is the minimum of the ratio

$$\frac{\sum_j a_{\alpha j} \bar{x}_j - b_\alpha}{- \sum_j a_{\alpha j} u_j}$$

for all α such that $\sum a_{\alpha j} u_j < 0$.
Therefore the vector u is attainable at $\bar{x}$.

II-13.3 Necessary Condition: Kuhn-Tucker Theorem for Problem P

E_{31} | If the function f has on F a local minimum at $\bar{x}$, if f is differentiable at $\bar{x}$ and if g satisfy the Kuhn-Tucker constraint qualification at $\bar{x}$, then there exists a vector $\bar{y}$ of R , such that

$$\bar{y} \geqslant 0 \ , \ \bar{y}^T g(\bar{x}) = 0 \ , \ f'(\bar{x}) - \bar{y}^T g'(\bar{x}) = 0$$

The above three relations are called the Kuhn-Tucker conditions.

ii) Geometrical interpretation. The Kuhn-Tucker conditions mean that $\nabla f(\bar{x})$ ($= (f'(\bar{x}))^T$) is a convex linear combination of the $\nabla g_i(\bar{x})$ for which $g_i(\bar{x}) = 0$.

iii) Proof. If $g_i(\bar{x}) > 0$ for all i, that is to say if $\bar{x}$ is in the interior of F, then

$f'(\bar{x}) = 0 \, (E_{19})$ and the theorem holds with $\bar{y} = 0$.

Otherwise, let u be an arbitrary vector locally constrained at $\bar{x}$, i.e.

$$(23) \qquad u^T \nabla g_i (\bar{x}) \geqslant 0 \quad \text{for all i of } N(\bar{x})$$

Then, by the Kuhn-Tucker qualification, the vector u is also attainable at $\bar{x}$. Let ψ (t) be a corresponding vector function. According to the differentiability properties of the hypothesis we have

$$\frac{df(\psi)}{dt} (0) = f'(\bar{x}) \frac{d\psi}{dt} (0)$$

Now, in a neighbourhood of $\bar{x}$ in F, $f(\bar{x})$ is smaller (in the wide sense) than $f(x)$. Hence, the first member is positive; so is the second one:

$$(24) \qquad f'(\bar{x})u \geqslant 0 \quad , \text{ or } \quad u^T \nabla f(\bar{x}) \geqslant 0$$

Thus all vectors u satisfying (23) also satisfy (24). By Farkas theorem ((15), chapter I) there exist numbers y_j such that

$$(25) \qquad y_i \geqslant 0 \quad , \quad \nabla f(\bar{x}) = \sum_{i \in N(\bar{x})} y_i \nabla g_i(\bar{x})$$

Let $\bar{y}$ be the m—vector constructed as follows: its components having numbers i of $N(\bar{x})$ are the previous $\bar{y}_i$, its other components are 0. Then, according to (25), we have

$$\bar{y} \geqslant 0 \qquad f'(\bar{x}) - \sum_{i=1}^{m} \bar{y}_i \, g_i' (\bar{x}) = 0$$

Moreover

$$\sum_{1}^{m} \bar{y}_i \, g_i (\bar{x}) = 0$$

because for each i either the first or the second factor is zero.

This ends the proof of the Kuhn-Tucker theorem for problem P.

iv) The Kuhn-Tucker theorem for problem P' is reduced to the previous one: we replace the equality $h(x) = 0$ by the inequalities $h(x) \geqslant 0$, $-h(x) \geqslant 0$, and we immediately obtain the following statement.

E_{32}

Kuhn-Tucker theorem for Problem P'

If the function f has a local minimum at $\bar{x}$, and if g and h satisfy the Kuhn-Tucker constraint qualification at $\bar{x}$, then there exist a vector $\bar{y}$ of R^m and a vector $\bar{z}$ of R^P such that

$$\bar{y} \geqslant 0 \ , \ \bar{y}^T g(\bar{x}) + \bar{z}^T h(\bar{x}) = 0 \ , \ f'(\bar{x}) - \bar{y}^T g'(\bar{x}) - \bar{z}^T h'(\bar{x}) = 0$$

II-14 Sufficient Condition: Converse of Kuhn-Tucker Theorem in Convex-Concave Optimization

Kuhn-Tucker conditions are, by themselves, not sufficient for a point to be optimal. But they become sufficient when some properties of convexity and concavity are prescribed.

E_{33}

Converse of Kuhn-Tucker Theorem for Problem P

We make the following hypothesis: the set S is open and convex, the function f is convex (strictly convex), the function g is concave. If there exists on F a point $\bar{x}$ where the functions f and g are differentiable and a vector $\bar{y}$ of R^m such that

$$\bar{y} \geqslant 0 \ , \ \bar{y}^T g(\bar{x}) = 0 \ , \ f'(\bar{x}) - \bar{y}^T g'(\bar{x}) = 0$$

then the function f has on F at $\bar{x}$ a minimum (a strict minimum).

Proof. The function

$$f(x) - \bar{y}^T g(x)$$

is convex on S. Its derivative is null at $\bar{x}$. Therefore, it has at this point a minimum (E_{23}):

$$f(\bar{x}) - \bar{y}^T g(\bar{x}) \leqslant f(x) - \bar{y}^T g(x) \ , \ \text{for each x}$$

of S, and, with greater reason, for each x of F.

Now, $\bar{y}^T g(\bar{x}) = 0$, and $\bar{y}^T g(x) \geqslant 0$ on F. Therefore

$$f(\bar{x}) \leqslant f(x) \quad \text{on} \quad F$$

Hence, the first part of the theorem (f convex) is proved.

For the second part (f strictly convex), the function $f(x) - \bar{y}^T g(x)$ is strictly convex. The proof is similar to the one of the first part by replacing wide inequalities by strict inequalities and "for each x" by "for each x different from $\bar{x}$".

II-15 Saddlepoint Sufficient Optimality Condition

This condition requires neither convexity or concavity nor differentiability assumptions.

II-15.1 Saddlepoint of the Lagrangian

E₃₄

> Definition of the Lagrangian L of Problem P
> It is the function L defined on $S \times (R^+)^m$ by
>
> $$(x,y) \longrightarrow L(x,y) = f(x) - y^T g(x)$$

E₃₅

> Definition of a Saddlepoint of Lagrangian L
> The point $(\bar{x},\bar{y})$ of $S \times (R^+)^m$ is a saddlepoint of L if and only if
>
> $$L(\bar{x},y) \leqslant L(\bar{x},\bar{y}) \leqslant L(x,\bar{y})$$
>
> for all (x,y) of $S \times (R^+)^m$.

II-15.2 Sufficient Optimality Condition

E₃₆

> If $(\bar{x},\bar{y})$ is a saddlepoint of the Lagrangian L of problem P, then $\bar{x}$ is a solution of problem P.

Proof. From the first inequality of the definition of L we have

$$(y - \bar{y})^T g(\bar{x}) \geqslant 0 \quad \text{for all } y \text{ in } (R^+)^m$$

For any j, $1 \leqslant j \leqslant m$, let

$$\begin{cases} y_j = \bar{y}_j + 1 \\ y_{j'} = \bar{y}_j \quad \text{for } j' = 1, \ldots j - 1, j + 1, \ldots m \end{cases}$$

It follows that $g_j(\bar{x}) \geqslant 0$. Hence $g(\bar{x}) \geqslant 0$ and $\bar{x}$ is a feasible point.

Now, in the first inequality of the definition of L, let us take $y = 0$. According to $\bar{y} \geqslant 0$ and $g(\bar{x}) \geqslant 0$ we obtain

$$\bar{y}^T g(\bar{x}) = 0$$

Then, from the second inequality of the definition of L, we get

$$f(\bar{x}) \leqslant f(x) - \bar{y}^T g(x) \qquad (26)$$

for all x in S and, with greater reason, for all x in F. But on F we have $g(x) \geqslant 0$. Then inequality (26) gives

$$f(\bar{x}) \leqslant f(x) \quad \text{for each } x \text{ in F}$$

Hence $\bar{x}$ is a solution of problem P.

II-16 Necessary Saddlepoint Condition in Convex Concave Optimization

Here, we shall need some properties of convexity, concavity and regularity.

E 37

Necessary Saddlepoint Optimality Condition

We make the following assumptions: S is convex, f is convex, g is concave, F has an interior point. Then, if $\bar{x}$ is a solution of problem P, there exists a vector $\bar{y}$ of $(R^+)^m$ such that $(\bar{x}, \bar{y})$ is a saddlepoint of the Lagrangian L of problem P.

Proof. Let us define the following subset A of R^{m+1} by

$$A = \{(z,\zeta) \mid \text{ there exists an x in S such that}$$

$$z \geqslant f(x) \, , \, \zeta \leqslant g(x)\}$$

By using the convexity of f and (– g) it is easy to show that A is convex. The point $(f(\bar{x}), 0)$ is a boundary point of A. Therefore A has a supporting plane at this point: there exists a non-null vector (λ – μ) such that

$$\lambda z - \mu^T \zeta \geqslant \lambda f(\bar{x})$$

for all (z, ζ) in A. Because A contains remote points in the directions (z, 0), (0, – ζ), we have $\lambda \geqslant 0$, $\mu \geqslant 0$.

The point $(f(x), g(x))$ is in A. Therefore

(27) $$\lambda f(\bar{x}) \leqslant \lambda f(x) - \mu^T g(x)$$

The fact that F has an interior point x^o, i.e.

(28) $$g(x^o) > 0$$

implies that $\lambda > 0$. Otherwise, according to (27) and to (λ, μ) $\neq 0$, we have $g(x) \leqslant 0$ for each x in F, which contradicts (28).

Let $\bar{y} = \mu/\lambda$. Then, inequality (27) becomes

(29) $$f(\bar{x}) \leqslant f(x) - \bar{y}^T g(x) \quad , \quad \bar{y} \geqslant 0$$

for all x of S.

Putting x = $\bar{x}$ in (29) we obtain $\bar{y}^T g(\bar{x}) \leqslant 0$. But we have $\bar{y} \geqslant 0$ and $g(\bar{x}) \geqslant 0$. Hence

(30) $$\bar{y}^T g(\bar{x}) = 0$$

and (29) is written as follows

(31) $$\begin{cases} f(\bar{x}) - \bar{y}^T g(\bar{x}) \leqslant f(x) - \bar{y}^T g(x) \\ \text{for all } x \text{ in S} \end{cases}$$

In another way, for every y of $(R^+)^m$ we have $y^T g(\bar{x}) \geqslant 0$ and consequently.

$$f(\bar{x}) - y^T g(\bar{x}) \leqslant f(\bar{x}) \tag{32}$$

Finally, inequalities (31) and (32) (with (30)) show that $(\bar{x}, \bar{y})$ is a saddlepoint of L.

II-17 Duality

II-17.1 Definition of the dual problem of P

First, we introduce some new notations.

E_{38}

> **Notations**
> i) θ, function defined on $(R^+)^m$ by
>
> $$\theta(y) = \inf_{x \in S} L(x,y)$$
>
> ii) $\Lambda = \{y \mid y \in (R^+)^m, \inf \theta(y) > -\infty\}$

Now, let us remark that

$$\sup_{y \in (R^+)^m} L(x,y) = \begin{cases} f(x) & \text{if } x \text{ is in } F \\ +\infty & \text{if } x \text{ is not in } F \end{cases}$$

Therefore, the problem P can be written

$$P \begin{cases} \min & \sup L(x,y) \\ x \in S, & y \in \Lambda \end{cases}$$

According to a general definition, the dual of P is obtained by inverting min (or inf) and sup (or max). More precisely, we state this dual problem D as follows

$$D \begin{cases} \max \inf L(x,y) \\ y \in \Lambda, & x \in S \end{cases}$$

E_{39} | The dual of P is the problem D: maximize θ on Λ.

II-17.2 Example: Linear Optimization

Let us return to problem P of chapter I:

$$P \{ \text{minimize } c^T x \text{ on } x \geqslant 0, \; Ax - b \geqslant 0$$

Here, the subset S of R^n is defined by $x \geqslant 0$. The Lagrangian is

$$c^T x - y^T (Ax - b) = (c^T - y^T A)x + y^T b$$

The function θ is defined by

$$\theta(y) = \begin{cases} y^T b \text{ if } c^T - y^T A \geqslant 0 \\ -\infty \text{ in the other cases} \end{cases}$$

According to E_{38} and E_{39}, the dual of P is

$$D \{ \text{maximize } y^T b \text{ on } y \geqslant 0, \; c^T - y^T A \geqslant 0$$

This is the problem you have defined in chapter I.

II-17.3 Relations Between Problems P and D (E_{39})

We deduce these relations from the saddlepoint theorem and from the following property

(33) | For each x in F and each y in $\wedge$ we have

$$\theta(y) \leqslant f(x)$$

Indeed $\theta(y) \leqslant f(x) - y^T g(x)$, and on F and $\wedge$ we have $g(x) \geqslant 0$ and $y^T \geqslant 0$. Therefore

$$\theta(y) \leqslant f(x)$$

i) The first relation is given by the theorem below.

E_{40} | If the point $\bar{x}$ of F and the point $\bar{y}$ of $\wedge$ satisfy the relation $\theta(\bar{y}) = f(\bar{x})$, they are respectively solutions of P and D.

Proof. Let x be an arbitrary point of F and let y be an arbitrary point of $\wedge$. According to property (33) and the hypothesis we have

$$f(\bar{x}) \leqslant f(x) \;,\; \theta(y) \leqslant \theta(\bar{y})$$

These two inequalities prove the theorem.

ii)

E41 | Let $\bar{x}$, $\bar{y}$ be two points of S, $(R^+)^m$ respectively. $(\bar{x}, \bar{y})$ is a saddlepoint of the Lagrangian L of problem P, if and only if $\bar{x}$, $\bar{y}$ is in F, $\wedge$ respectively and $\theta(\bar{y}) = f(\bar{x})$.

Proof of the necessity. In (II–15.2) we have shown that $\bar{x}$ is in F and that $\bar{y}^T g(\bar{x}) = 0$. This equality implies $L(\bar{x}, \bar{y}) = f(\bar{x})$. The second inequality of the definition of L gives $L(\bar{x}, \bar{y}) = \theta(\bar{y})$, and shows that $\bar{y}$ is in $\wedge$.

Proof of the sufficiency. By the definition of $\bar{y}$ and the hypothesis we have

$$\theta(\bar{y}) = f(\bar{x}) \leqslant f(\bar{x}) - \bar{y}^T g(\bar{x})$$

and consequently $\bar{y}^T g(\bar{x}) \leqslant 0$.

But the product $\bar{y}^T g(\bar{x})$ is positive. Therefore

$$\bar{y}^T g(\bar{x}) = 0$$

On the other hand, $\theta(\bar{y})$ is the infimum of $L(x, \bar{y})$ when x describes S:

$$f(\bar{x}) - \bar{y}^T g(\bar{x}) \leqslant f(x) - \bar{y}^T g(x)$$

Hence, the two inequalities of the definition of L hold.

iii) Thus far we have made no assumption of convexity, concavity or regularity. This is not the case for the following theorem.

E42 | Let us suppose that f is convex, g is concave, and F has an interior point. If the problem P has a solution $\bar{x}$, then the problem D has also a solution $\bar{y}$ and $\theta(\bar{y}) = f(\bar{x})$.

Indeed we successively have: there exists $\bar{y}$ in $(R^+)^m$ such that $(\bar{x}, \bar{y})$ is a saddlepoint of $L(E_{37})$, $\bar{y}$ is in $\wedge$ and $\theta(\bar{y}) = f(\bar{x})$, $\bar{y}$ is a solution of $D(E_{41})$.

F. APPLICATIONS TO MECHANICAL PROBLEMS

II-18 Optimization of Vibrating Structures

Let F be the subset of the vectors x of R^n such that $x > 0$, and let S be a family of structures. We suppose that to each x of F corresponds a structure S(x) of S and conversely, and that the generalized mass matrix, the generalized stiffness matrix and the mass of S(x) are respectively

$$(34) \qquad A(x) = A_o + \sum_1^n x_j A_j \quad ,$$

$$(35) \qquad C(x) = C_o + \sum_1^n x_j C_j \quad ,$$

$$(36) \qquad m(x) = m_o + \sum_1^n x_j \lambda_j \quad ,$$

where A_o, C_o, A_j, C_j are given symmetrical (n, n)—matrices, with A_o and C_o semidefinite positive, the A_j's and the C_j's definite positive, and where m_o is positive and the λ_j's are strictly positive.

Let $\omega(x)$ be the fundamental frequency (pulsation) of $S(x)$ and let $u(x)$ be a corresponding eigenvector. We suppose that, for each x, the u's are normed by

$$(37) \qquad u^T Au = 1$$

According to the definition we have

$$(38) \qquad (C - \omega^2 A)u = 0$$

(*) See also – TURNER (M.J.), Design of minimum mass structures with specified natural frequencies, AIAA Journal, March 1967, p. 406–412.

ZARGHAMEE, Optimal frequency of structures, AIAA Journal, April 1968, p. 749–750.

In order to avoid possible resonances, it is to be wished that the fundamental frequency be maximum. Then, we consider the two following problems where the design variables are the x_j's:

$$
\begin{cases}
\text{\underline{Problem 1} $-$ \underline{The mass $m(x)$ being given, maximize the fundamental frequency}} \\
\text{\underline{$\omega(x)$.}} \qquad\qquad\qquad\qquad\qquad\qquad\qquad\qquad\qquad\qquad\qquad\qquad\qquad\qquad (39) \\
\text{\underline{Problem 2} $-$ \underline{The fundamental frequency $\omega(x)$ being given, minimize the mass}} \\
\text{\underline{$m(x)$.}}
\end{cases}
$$

II-18.1 <u>Necessary condition.</u> We are going to show that it is possible to calculate the derivative $d(\omega^2)/dx_j$, so that Euler method with multiplier is well fitted. Indeed, from (38) we have

$$
(C_j - \omega^2 A_j)\, u - \frac{d(\omega^2)}{dx_j}\, Au + (C - \omega^2 A)\, \frac{du}{dx_j} = 0
$$

We multiply by u^T on the left. The factor $u^T(C - \omega^2 A)$ appears in the last term; according to (38) this factor is null. Taking (37) into account, we obtain

$$
\frac{d(\omega^2)}{dx_j} = u^T (C_j - \omega^2 A_j)\, u \qquad\qquad\qquad (40)
$$

Then, theorem (II-12) with a multiplier can be applied: <u>In order that a vector x of F be solution of problem 1 or problem 2 it is necessary that it verifies the following condition</u>

$$
\frac{u^T(x)[\, C_j - \omega^2(x)\, A_j\,]\, u(x)}{\lambda_j} = H,\ \text{const. independent of } j \qquad (41)
$$

II-18.2 <u>Sufficient condition.</u>

The fundamental frequency of $S(x)$ is given by the Rayleigh ratio,

$$
\omega^2(x) = \frac{u^T(x)C_o u(x) + \sum_1^n x_j u^T(x)C_j u(x)}{u^T(x)A_o u(x) + \sum_1^n x_j u^T(x)A_j u(x)}
$$

For an arbitrary vector x^* of F and the corresponding structure $S(x^*)$, the Rayleigh ratio is minimum for the fundamental eigenvectors. Therefore we have

$$
\omega^2(x^*) \leqslant \frac{u^T(x)C_o u(x) + \sum_1^n x_j^* u^T(x)C_j u(x)}{u^T(x)A_o u(x) + \sum_1^n x_j^* u^T(x)A_j u(x)}
$$

From these two relations we deduce by substraction

(42)
$$(\omega^2(x^*) - \omega^2(x))[u^T(x)A_o u(x) + \sum_1^n x_j^* u^T(x)A_j u(x)]$$
$$\leqslant \sum_1^n (x_j^* - x_j)[u^T(x)C_j u(x) - \omega^2(x)u^T(x)A_j u(x)]$$

In the left-hand side of (42), the factor in square brackets, noted $\phi(x^*, x)$, is strictly positive.

Let us suppose that the necessary condition (41) is verified. Then inequality (42) gives

(43) $$(\omega^2(x^*) - \omega^2(x))\ \phi(x^*,x)\ \leqslant\ (m(x^*) - m(x))H.$$

— For problem 1, the mass is given; relation (43) shows that $\omega(x) \geqslant \omega(x^*)$ for every vector x^* of F that is for every structure of the family S: <u>In order that the vector x of F be solution of problem 1 it is sufficient that it verifies condition</u> (41).

— For problem 2, the fundamental frequency is given. The constant H is strictly positive, then relation (43) shows that $m(x) \leqslant m(x^*)$ for every vector x^* of F that is for every structure of the family S: <u>In order that the vector x of F be solution of problem 2 it is sufficient that it verifies condition</u> (41).

II-19 Optimization of Some Elastic Beams
II-19.1 <u>Presentation</u>

In this paragraph (II-19) we consider a family B of beams with given length called ℓ, given loads and given end conditions (these conditions being isostatic). We suppose that the stiffness density s of each beam of B is piecewise continuous and conversely that every function s defined on $[\ 0, \ell\ ]$, positive and piecewise continuous is the stiffness density of a unique beam of B. We call B(s) the beam whose stiffness density is the function s. Let μ (s, x) be the mass density along the beam B(s) at the point x.

The functions s and μ are assumed to verify the following given equality

(44) $$\mu(s,x) = a(x)\ s^k + \alpha(x)$$

where k is a strictly positive number, where a(x), α (x) are given piecewise continuous positive functions, a(x) being strictly positive.

There are numerous classical examples where such a relation exists. Let us give some of them (E, ρ are respectively the Young modulus and the volumic mass).

— First, let us consider rectangular homogeneous cross sections of breadth b and height h.

If h is a given constant and if b is an unknown function, equality (44) holds with

$$k = 1 \ , \quad a = \frac{12\rho}{Eh^2} \ , \quad \alpha = 0$$

On the contrary, if b is a given constant and if h is an unknown function, we also have (44) with

$$k = \frac{1}{3} \ , \quad a = \rho \left(\frac{12b^2}{E} \right)^{\frac{1}{3}} , \quad \alpha = 0$$

More generally, if h has the form h = b^ν where ν is a given strictly positive constant, then

$$k = \frac{1 + \nu}{3 + \nu} \ , \quad a = \rho \left(\frac{12}{E} \right)^{\frac{1 + \nu}{3 + \nu}} , \quad \alpha = 0$$

— Now, let us consider sandwich beams where the common thickness of upper and lower face sheets is assumed to be small in comparison to the height h. Then, if b is the breadth of the core, if ρ_1 and ρ_2 are respectively the volumic masses of the core and of the face sheets, we have

$$k = 1, \quad a = \frac{4\rho_2}{Eh^2} \ , \quad \alpha = \frac{bh(3\rho_1 - \rho_2)}{3}$$

— Finally, suppose that the beams B are thin walled tubes (radius r, thickness e).

If e is given, we have

$$k = \frac{1}{3} \ , \quad a = 2\rho \left(\frac{\pi^2 e^2}{E} \right)^{\frac{1}{3}} , \quad \alpha = 0$$

If r is given, we have

$$k = 1 \quad, \quad a = \frac{2\rho}{Er^2} \quad, \quad \alpha = 0$$

Of course, in the previous expressions of a and α, the b, h, r, e's are given strictly positive piecewise continuous functions (for example, continuous or piecewise constant).

Now, let us return to the case of the general formula (44).

The mass of the beam B(s) is

$$(45) \qquad\qquad m(s) = \int_0^\ell \mu[s(x), x]\, dx$$

On the other hand, let us consider the deflection of beams B at a given point $\xi, 0 \leqslant \xi \leqslant \ell$. According to the principle of virtual work, the deflection of the beam B(s) at this point is

$$(46) \qquad\qquad z(s) = \int_0^\ell \frac{P(x)}{s(x)}\, dx$$

with
$$(47) \qquad\qquad P(x) = M(x)\, M_1(x)$$

where M(x) is the bending moment at ξ and where $M_1(x)$ is the bending moment at the same point when the given loads are replaced by the unit load applied to the point ξ (Note that M and M_1 are independent of the function s, since beams B are isostatic).

It is often of interest to solve the two following problems when the design variable is the stiffness function s.

$$(48) \begin{cases} \text{Problem 1 – \underline{The mass} m(s) \underline{being given, minimize the deflection} z(s).} \\ \text{Problem 2 – \underline{The deflection} z(s) \underline{being given, minimize the mass} m(s).} \end{cases}$$

It was pointed out by (R.L.) BARNETT * that if there are points where P is strictly positive and points where it is strictly negative, then problem 2 has no

(*) BARNETT (R.L.), Minimum weight design of beams for deflection, Journal of the Engineering Mechanics Division, February 1961, p. 75–109.

solution. We shall treat the more interesting case where P(x) keeps a constant sign, for example the sign +.

II-19.2 Necessary Condition

Theorem (II-12) gives a necessary condition of optimality for both problem 1 and problem 2, with one multiplier y. The corresponding Euler equation is

$$yP = as^{k+1} \tag{49}$$

The value of this multiplier y is obtained by (49) and the constraint equality. Now, (49) gives optimal values of s. The results are the following ones.

Problem 1. If $\bar{s}$ is a solution of problem 1 (m(s) = m_0, given), then

$$\bar{s} = \left(\frac{P}{a}\right)^{\frac{1}{k+1}} m_0^{\frac{1}{k}} \left\{ \int_0^\ell P^{\frac{k}{k+1}} a^{\frac{1}{k+1}} dx \right\}^{-\frac{1}{k}} \tag{50}$$

and the minimum value of z(s) is

$$\bar{z} = m_0^{-\frac{1}{k}} \left\{ \int_0^\ell P^{\frac{k}{k+1}} a^{\frac{1}{k+1}} dx \right\}^{\frac{k+1}{k}} \tag{51}$$

Problem 2. If $\bar{s}$ is a solution of problem 2 (z (s) = z_0, given), then

$$\bar{s} = \left(\frac{P}{a}\right)^{\frac{1}{k+1}} z_0^{-1} \int_0^\ell P^{\frac{k}{k+1}} a^{\frac{1}{k+1}} dx \tag{52}$$

and the minimum value of m(s) is

$$\bar{m} = z_0^{-k} \left\{ \int_0^\ell P^{\frac{k}{k+1}} a^{\frac{1}{k+1}} dx \right\}^{k+1} \tag{53}$$

II-19.3 Sufficient Condition

The following methods use a Hölder inequality; an analogous method was given by R.L. BARNETT*

_____

(*) loc. cit.

For problem 1, we have

$$
m_o^{\frac{1}{k}} \left[z(s) - z(\bar{s}) \right] = \int_o^\ell \frac{P}{s}\, dx \left(\int_o^\ell as^k dx \right)^{\frac{1}{k}} - \left\{ \int_o^\ell \left(\frac{P}{s} \right)^{\frac{k}{k+1}} \left(as^k \right)^{\frac{1}{k+1}} dx \right\}^{\frac{k+1}{k}}
$$

so that the difference $z(s) - z(\bar{s})$ has the same sign as

$$
(54) \qquad \left(\int_o^\ell \frac{P}{s}\, dx \right)^{\frac{k}{k+1}} \left(\int_o^\ell as^k\, dx \right)^{\frac{1}{k+1}} - \int_o^\ell \left(\frac{P}{s} \right)^{\frac{k}{k+1}} \left(as^k \right)^{\frac{1}{k+1}} dx
$$

But (54) is positive by Hölder inequality. Then

$$
z(\bar{s}) \leqslant z(s) \quad \text{for all } s .
$$

For problem 2, we have

$$
y_o^{\frac{k}{}} \left[m(s) - m(\bar{s}) \right] = \left(\int_o^\ell \frac{P}{s}\, dx \right)^k \int_o^\ell as^k dx - \left\{ \int_o^\ell \left(\frac{P}{s} \right)^{\frac{k}{k+1}} \left(as^k \right)^{\frac{1}{k+1}} dx \right\}^{k+1}
$$

so that the difference $m(s) - m(\bar{s})$ has the same sign as (54). Therefore we have

$$
m(\bar{s}) \leqslant m(s) \quad \text{for all } s
$$

Stiffness $\bar{s}$ given by (50) and (52) are respectively solutions of problem 1 and problem 2.

II-20 Optimization of the Velocity of a Spatial Vehicle

Let us consider flights of a spatial vehicle in a uniform and constant gravitational field (acceleration $\vec{g}$) with the following assumptions and data: the initial time $t_o = 0$, the initial position O, the initial velocity $\vec{V}_o$, the initial mass m_o are given; the trajectories lie in a given plane (containing O, $\vec{g}$, $\vec{V}_o$); the coordinate

basis $\vec{x}$, $\vec{y}$ of this plane is such that $\vec{g} = g\vec{y}$ $(g > 0)$ and $\vec{V}_0 = v_0 \cos \alpha \, \vec{x} +$
$+ \, v_0 \sin \alpha, \vec{y}$ $(v_0 > 0, -\pi/2 < \alpha < \pi/2)$; the scalar velocity of gas, the mass and its variation rate are respectively called $u(t)$, $m(t)$, $\beta(t)$ (which are given piecewise continuous positive functions). It is assumed that the mission which is explained below is trusted to the propeller and that the corresponding final mass m_1 is given.

The mass at the time t is

$$m(t) = m_0 - \int_0^t \beta(\theta)\,d\theta$$

so that the time t_1 of the end of the mission is known.

The scalar thrust acceleration

$$\gamma = \frac{c\beta}{m}$$

is a known function of t. On the contrary, the unit vector $\vec{e}$ of the thrust is an unknown function of t. The following problem is met when launching satellites.

(55)
> How to choose the direction $\vec{e}$ (t) of the thrust, that is to say the angle
> $$\varphi(t) = (\vec{x}, \vec{e}(t)) \,, \; -\frac{\pi}{2} < \varphi(t) < \frac{\pi}{2}$$
> in such a way that, at the time t_1 the velocity vector $\vec{V}$ be parallel to $\vec{x}$ and have a maximum intensity?

The equation of the motion

$$\frac{d\vec{V}}{dt} = \gamma(t) \, \vec{e} + \vec{g}$$

gives the following components of $\vec{V}$ at the time t_1 :

$$\dot{x}(t_1) = \int_0^{t_1} \gamma(t) \cos \varphi(t)\,dt + v_0 \cos \alpha$$

$$\dot{y}(t_1) = \int_0^{t_1} \gamma(t) \sin \varphi(t) dt - g\, t_1 + v_0 \sin \alpha$$

We have in view an application of theorem (II-12) about Lagrange multipliers. Here, the normed space ϵ is the space of functions φ of C^1 class defined on $[0, t_1]$, S is the subspace of ϵ defined by $-\pi/2 < \varphi(t) < \pi/2$, so that the optimal problem (55) becomes

(56)

$$
\begin{cases}
\text{Maximize} \\[4pt]
\qquad f(\varphi) = \int_0^{t_1} \gamma(t) \cos \varphi(t) dt \\[12pt]
\text{on the subset } F \text{ of } S \text{ defined by} \\[4pt]
\qquad \int_0^{t_1} \gamma(t) \sin \varphi(t)\, dt = g t_1 - v_0 \sin \alpha
\end{cases}
$$

Writing Euler equation with one multiplier and calculating this multiplier by means of the constraint, we obtain the following result: If problem (55) has a solution, this solution is constant and defined by

(57)
$$\sin \varphi(t) = \frac{g t_1 - v_0 \sin \alpha}{\int_0^{t_1} \gamma(t)\, dt}$$

It is proved, by optimal control theory, that $\varphi(t)$ (57) is really the solution of problem (55).

REFERENCES

[1]   ABADIE J.: *Nonlinear Programming,* North Holland, 1967.

[2]   ARROW K.J., HURWICZ L., UZAWA H.: *Studies in Linear and Nonlinear Programming,* Stanford University Press, 1958.

[3]   BROUSSE P.: Optimisation des structures mécaniques, *Les méthodes d'optimisation dans la construction,* Séminaire du Collège International des Sciences de la Construction, Saint-Rémy-lès-Chevreuse, France, 6-9 novembre 1973.

[4]   HADLEY G.: *Nonlinear and dynamic programming,* Addison-Wesley, 1964.

[5]   MANGASARIAN O.L.: *Nonlinear Programming,* McGraw-Hill, 1969.

[6]   PRAGER W.: Optimality criteria derived from classical extremum principles, An introduction to structural optimization, University of Waterloo, 1968.

[7]   VAJDA S.: Tests of Optimality in Constrained Optimisation, J. Inst. Math. Applic., 13, p. 187-200, 1974.

CHAPTER III

SOME NUMERICAL METHODS

Actually, it is not possible to work in the field of structural optimization without using computers. Hence, the numerical methods have become essential. These numerical methods are very numerous and we cannot make a full study of them. We shall be content with a few extremely succinct general ideas.

A. PROBLEMS WITHOUT CONSTRAINTS

The problem we are concerned with is stated as follows (compare to Chap. II, E_{18}).

$$E_1 \left| \begin{array}{l} \text{Minimize a differentiable (possibly twice differentiable) numerical} \\ \text{function } f \text{ defined on a given open set in } R^n . \end{array} \right.$$

III-1. Iterative Methods

Let us suppose that the given function f has a strict minimum at $\bar{x}$. In order to compute an approximate value of $\bar{x}$ we perform by iteration: given the kth approximation x^k, we define the next one, i.e. x^{k+1}, by

$$(1) \qquad x^{k+1} = x^k + \lambda^k w^k$$

where w^k is a non null vector and λ^k a positive ($\geqslant 0$) or negative ($\leqslant 0$) number w^k and λ^k being suitably chosen. Hence, there are two local problems: choice of direction w^k, choice of number λ^k. There are numerous convergence criteria, but empirical processes are often used.

Let f be differentiable. Then it has the following development

$$(2) \qquad f(x) = f(x^k) + (\nabla f(x^k))^T (x - x^k) + \epsilon_k (x) \| x - x_k \|$$

where $\epsilon_k (x)$ tends to zero when x tends to x^k.

Let us suppose that $\nabla f(x^k) \neq 0$. Since we are looking for the minimum of f, it is reasonable to try to reduce $f(x)$ from $f(x^k)$. Saying that, at a point x infinitely near to x^k, we have: $f(x) < f(x^k)$ is equivalent to $(\nabla f(x^k))^T (x - x^k) < 0$, or to the angle between the vectors $\nabla f(x^k)$ and $(x - x^k)$ is obtuse. If such is the case, the direction $(x - x^k)$ is said to be a <u>descent direction</u>. Generally, the vector w^k is chosen in a descent direction. Given the point x^k and the descent direction w^k , we consider the following function of ρ defined by

$$\varphi_k (\rho) = f(x^k + \rho w^k) \ , \ \rho \geqslant 0 \tag{3}$$

and the choice of ρ remains to be done.

Fairly often, ρ is taken in such a way that the function φ_k is minimal. Then, one says that ρ is chosen <u>in optimal manner</u>.

If that is the case, a necessary optimality condition is written as follows: the derivative of φ_k is null at ρ^k, i.e.

$$(\nabla f(x^{k+1}))^T \ w^k = 0 \tag{4}$$

Another method consists in using an approximation or an over-estimation of φ_k . For instance, let f verify

$$f(x^k + \rho w^k) \leqslant f(x^k) + \rho (\nabla f(x^k))^T w^k + \frac{1}{2} \rho^2 M \| w^k \|^2$$

where M is a strictly positive constant number. Often, ρ^k is taken as the value of ρ for which the right hand side is minimum. At other times, a quadratic or a cubic approximation is used.

III-2 The Steepest Descent

This method consists in taking

$$w^k = - \nabla f(x^k) \tag{5}$$

and in choosing ρ in optimal manner.

It follows from (4) that successive gradients in the procedure are orthogonal, that is, for all k

$$(\nabla f(x^{k+1}))^T \ \nabla f(x^k) = 0 \tag{6}$$

It can be proved that any limit point $\bar{x}$ of the sequence $\{ x^k \}$ is a stationary point, that is $\nabla f(\bar{x}) = 0$.

Unfortunately, "steepest descent" has several disadvantages:

- It is not generally a finite procedure for minimizing a positive definite quadratic function (and such functions are very important in mechanics!).
- Each approximation is computed independently of the others; that is, no information is stored and used that might accelerate convergence.
- The rate of convergence depends strongly on the graph of the function. In many cases, steepest descent generates short zigzagging moves in a neighbourhood of any solution.

III-3 Conjugate Direction Methods

III-3.1 Introduction

Let f be twice differentiable. Then near any point x^o the following development holds

$$
\begin{aligned}
f(x) = f(x^o) + (\nabla f(x^o))^T (x - x^o) + \\
+ \frac{1}{2} (x - x^o)^T Hf(x^o)(x - x^o) + \epsilon(x)\|x - x^o\|^2
\end{aligned}
$$

(7)

where $Hf(x^o)$ is the Hessian of the function f at the point x^o, and where $\epsilon(x)$ tends to zero when x tends to x^o.

Let us suppose that x^o is near the solution $\bar{x}$ of the optimization problem E_1 (we suppose that this solution $\bar{x}$ exists). Then, the matrix $Hf(x^o)$ is generally positive definite, and it is seen that, for numerous functions, the quadratic form

$$
(8) \qquad f(x^o) + (\nabla f(x^o))^T (x - x^o) + (x - x^o)^T Hf(x^o)(x - x^o)
$$

is a very good approximation of $f(x)$ in the proximity of x^o, and, consequently, of $\bar{x}$.

The importance of this general case on the one hand, and of the minimization problem of quadratic functions on the other hand, brings us along to turn our attention to the case where f is a quadratic function, noted q, i.e.

$$
(9) \qquad \begin{cases} q(x) = a + b^T x + \frac{1}{2} x^T Ax \\ \text{where A is a } \underline{\text{positive definite matrix}}. \end{cases}
$$

Let us note that the gradient and the Hessian of q are

$$\nabla q(x) = Ax + b \quad , \quad Hq(x) = A$$

We have proved that the function (9) is strictly convex (Chap. II, II-3 and E_8). Therefore it has a strict minimum (Chap. II, E_{23}) at the point $\bar{x}$ defined by

$$\nabla F(\bar{x}) = 0$$

that is to say

$$\bar{x} = -A^{-1}b \tag{10}$$

Unfortunately, such a method applied to a general function requires the knowledge of the second derivatives. This is a practically insurmountable difficulty. The so-called conjugate direction methods permit the computation of the minimum when the second derivatives are not available.

III-3.2 Optimization Via Conjugate Directions

First, let us recall some properties of quadratic functions q(9).

i) A set of n <u>non null</u> vectors p^k, k = 1, ...n, is <u>conjugate</u> with respect to the positive definite (n, n)-matrix A if and only if

$$p^{kT}Ap^j = 0 \quad \text{for all} \quad j \neq k \tag{11}$$

This definition also implies that the vectors p^k are linearly independent. Indeed, if

$$\sum_k \mu_k p^k = 0 \quad , \text{ then}$$

$$\sum_k \mu_k p^{jT} Ap^k = \mu_j p^{jT} Ap^j = 0$$

and $\mu_j = 0$ since A is positive definite.

ii) Given an approximation x^k of the solution $\bar{x}$ (10) and a search direction $w^k = p^k$, the next approximation x^{k+1} is

$$x^{k+1} = x^k + \lambda^k p^k$$

where λ^k is chosen to minimize $q(x^k + \lambda p^k)$ with respect to λ (positive or negative). Let us note that

$$q(x^k + \lambda p^k) = q(x^k) + \lambda p^{kT} \nabla q(x^k) + \frac{1}{2}(\lambda)^2 p^{kT} A p^k$$

and therefore that

$$(12) \qquad \lambda^k = -\frac{p^{kT} \nabla q(x^k)}{p^{kT} A p^k}$$

It should be remarked that the point x^{k+1} is the middle of the segment joining x^k to the second intersection of the straight line (x^k, p^k) with the ellipsoid $q(x) = q(x^k)$, and that the previous straight line is tangent at x^{k+1} to the ellipsoid $q(x) = q(x^{k+1})$.

Writing that the derivative of $q(x^k + \lambda p^k)$ with respect to λ is zero, we obtain a necessary and sufficient condition of optimality in the form

$$(13) \qquad (\nabla q(x^{k+1}))^T p^k = 0$$

From (12), we deduce that, if x^j is the solution $\bar{x}$, then all the next approximations x^{j+h}, $h \geq 0$, are stationary on $\bar{x}$.

Now, we are going to prove the following fundamental theorem.

(1) { The minimum is found in at most n iterations

Proof. Let us write successively

$$x^{k+2} = x^{k+1} + \lambda^{k+1} p^{k+1}$$

$$- - - - - - - - - - - -$$

$$x^{n+1} = x^n + \lambda^n p^n$$

$$(15) \qquad x^{n+1} = x^{k+1} + \sum_{k+1}^{n} \lambda^j p^j$$

We multiply this last equality by $p^{kT} A$ on the left. According to (13) and (11) we obtain

$$p^{kT} \nabla q(x^{n+1}) = 0$$

Now, the p^k's form a basis of R^n. Then

$$\nabla q(x^{n+1}) = 0$$

This implies

$$x^{n+1} = \bar{x}$$

Most of conjugate direction methods rely upon this theorem. The differences between the various methods are in the ingenuity by which conjugate directions are obtained and in the appropriateness of the iterative process to classes of functions to be minimized.

III-3.3 Example: The Davidon-Fletcher-Powell Method

Let us return to a general function f and to its development (7), and let us provisionally suppose that ρ is positive or negative. The D.F.P. algorithm is as follows

— Let $\{B_k\}$ be the sequence defined by
  · B_1 arbitrary positive definite (n, n)— matrix (for instance the unit matrix)

$$\cdot \quad B_{k+1} = B_k + \frac{\Delta_x^k \Delta_x^{kT}}{\Delta_x^{kT} \Delta g^k} - \frac{B_k \Delta g^k \Delta g^{kT} B_k}{\Delta g^{kT} B_k \Delta g^k}$$

with $\Delta_x^k = x^{k+1} - x^k$ and $\Delta g^k = \nabla f(x^{k+1}) - \nabla f(x^k)$
— Let us take

$$\cdot \quad w^k = - B_k \nabla f(x^k)$$

· λ^k chosen in optimal manner.

Then it can be proved that

i) B_k is a positive definite (n, n)—matrix for all k,

ii) the direction (x^k, w^k) is a descent direction,

iii) if the function to be minimized is quadratic (function a. (9)), the vectors

w^1, ... w^n are conjugate with respect to the matrix A,

iv) when f is an arbitrary function having the development (7), then B_k tends to $H^{-1}f(\bar{x})$ when k tends to infinity,

v) in the general case, and with some suitable assumptions, the process is convergent.

The D.F.P. method is fastened to the general "variable metric" method.

B. PROBLEMS WITH CONSTRAINTS

Without loss of generality we can only examine the inequality constraints case. So, the problem we are concerned with is stated as follows (compare to Chap. II, E_{27}).

E_2 | Let f, g_i (i = 1, ... m) be (m + 1) <u>differentiable numerical functions</u> <u>defined on an open set</u> Ω <u>in R^n and let g be the column matrix</u> $\{g_i\}$. <u>Minimize f(x)</u> <u>on the subset F of</u> Ω <u>defined by</u> $g(x) \geqslant 0$

We are only going to give a shorter presentation of the principle of a few classes of methods.

III-4 Approximation by a Sequence of Linear Problems

Let us suppose that f and g are defined on R^n.

We shall give a short description of the algorithm.

— Start from a point x^0 in F. Replace g(x) by its first order development around x^0 i.e.,

$$g^1(x) \;=\; g(x^0) \;+\; (\nabla g(x^0))^T (x - x^0)$$

and F by $F^1 : g^1(x) \geqslant 0$

In the same way, replace f(x) by its own first order development around x^0; this linear development is noted $f^1(x)$.

Finally, replace the problem E_2 by the following linear problem

| minimize $f^1(x)$ on F^1

Let x^1 be a solution of this problem (if such a solution exists).

— Operate from x^1 as we have operated from x^0, by linearizing f and g around x^1. Let x^2 be a solution of the new linear problem which have just been obtained. And so on.

Under some hard assumptions, principally assumptions of convexity, it can be proved that the sequence x^k tends to a solution of E_2 when k tends to infinity.

The main advantage of this method is that every such problem is a linear problem.

III-5. Penalty Methods

The penalty methods transform any problem with constraints into a problem without constraints.

III-5.1. General Method [1]

The open subset of Ω defined by $g(x) > 0$ is noted F^o.

Let us choose

i) a numerical function J of the n variables x_j satisfying the following two properties:

· J is defined and continuous on F^o

· for any point x^* such that $g_i(x^*) = 0$

for at least one i, and for any infinite sequence $\{x^k\}$ of points in F^o converging to x^*, we have $\lim_{k \to \infty} J(x^k) = + \infty$

ii) a numerical function s(r) of a strictly positive single variable r, satisfying the following three properties

· $s(r) > 0$ for all $r > 0$

· when r is decreasing, then s(r) strictly decreases

· for any sequence $r^k (r^k > 0)$ converging to zero, we have $\lim s(r^k) = 0$.

By definition, the function

$$U(x,r) = f(x) + s(r) J(x) \qquad (16)$$

is said to be a penalty function.

Then, the corresponding penalty technique proceeds as follows.

— Take $U(x, r^1)$ where $r^1 > 0$ is given. Suppose $U(x, r^1)$ has a local minimum (at a point noted $\bar{x}^1$) on F. The point $\bar{x}^1$ is in F^o, otherwise

---

(1) See FIACCO, A.V., and McCORMICK, G.P., Nonlinear Programming: Sequential Unconstrained Minimization Techniques, 1968.

$U(\bar{x}^1, r_1) = +\infty$, contradicting the supposition that U took a local minimum at $\bar{x}$. In other words, we have considered the following <u>unconstrained problem</u>: minimize $U(x, r^1)$ on the <u>open</u> set F^o in R^n. Of course, in order to compute $\bar{x}^1$, we start from a point $\bar{x}^o$ in F^o.

— Starting from $\bar{x}^1$, or from an approximate value $(\bar{x}^1)'$ of $\bar{x}^1$, in F^o and given r^2 such that $r^2 < r^1$, consider the minimization of $U(x, r^2)$ on F^o. Let $\bar{x}^2$ be a point (if it exists) where $U(x, r^2)$ has a local minimum.

— Continuing in this manner, start from $(\bar{x}^{k-1})'$ and find $\bar{x}^k$ where the function $U(x, r^k)$ has a local minimum.

Under appropriate assumptions, it can be proved that, when the sequence $\bar{x}^k$ exists, then it onverges to a local solution of the problem E_2.

It should be remarked that all the point $\bar{x}^k$ and their approximate values $(\bar{x}^k)'$ are in the interior F^o of F. This remark justifies the word "interior" in this class of penalty methods.

There are other kinds of penalty methods with other sorts of functions U, where some $\bar{x}^k$ are in the exterior of F. But, only the interior point minimization techniques are usable in mechanics. Indeed, every approximate solution should be in the feasible region.

III–5.2 <u>Examples</u>: Logarithmic Penalty Function, Fiacco–McCormick[1] Penalty Function

These functions are respectively

(17)
$$U(x, r) = f(x) + r \sum_1^m \log \frac{1}{g_i(x)}$$

(18)
$$U(x, r) = f(x) + r^2 \sum_1^m \frac{1}{g_i(x)}$$

To illustrate methods, we consider a very simple example

(19)
$$\begin{vmatrix} \text{minimize} & f(x) = x_1 + x_2 + x_2^2 \\ \text{on} & \begin{pmatrix} x_1 \\ x_2 \end{pmatrix} \geqslant 0 \end{vmatrix}$$

---

(1) FIACCO, A.V., and McCORMICK, G.P., Loc. cit.

Let us note that the solution of this problem is evident: it is x = 0.

In the two cases, the condition $\partial U/\partial x = 0$ is a necessary (and sufficient) minimality condition for U. Thus, we have the following components of the point $\bar{x}(r)$ where U(x, r) is minimum with respect to x:

$$\bar{x}(r) = \left(\frac{r}{\dfrac{-1 + \sqrt{1 + 8r}}{2}} \right)$$

$$\bar{x}(r) = \left(\begin{matrix} \sqrt{r} \\ \bar{x}_2 \end{matrix} \text{ ,unique solution of } 2x_2^3 + x_2^2 - r^2 = 0 \right)$$

We can effectively see that $\bar{x}(r)$ tends to the solution 0 when r tends to 0. It is easy to draw the trajectory of the point $\bar{x}(r)$. The numerical computation gives points $\bar{x}^1$, $\bar{x}^2$ ... lying on this trajectory and nearing more and more to zero.

III–6 The Gradient Projection Method [i]

We shall succintly explain the simplest algorithm, the one with linear constraints.

Let us consider, for example, the following problem where f is defined on R^n.

$$E_2' \quad \left| \begin{matrix} \underline{\text{minimize }} f(x), \ \underline{\text{convex, on }} g(x) \geqslant 0, \\ g_i(x) = a^{iT} x - b_i \ ; \ i = 1, \ldots m \end{matrix} \right.$$

III–6.1

Let x* be a feasible point (g (x*) $\geqslant$ 0). In order that x* be a solution of problem (E$'_2$), it is necessary and sufficient that

· $\nabla f(x^*) = 0$, if x* is an interior point (Chap. II, E_{23}),
· $\nabla f(x^*) - \Sigma \, y_i \, \nabla g \, (x^*) = 0$, i in $N(\bar{x})$, $y_i \geqslant 0$, if $N(\bar{x})$ is not empty, that is to say if there is at least one i such that $g_i(x^*) = 0$ (Chap. II, E_{33}).

III-6.2

Now, the gradient projection algorithm is as follows.

Let P_q be the projection-matrix on the subspace V_q spanned by the q first boundary planes. It is always possible, without loss of generality, to suppose that

i) See ROSEN, J.B., The Gradient Projection Method for Nonlinear Programming, S I A M Journal on Applied Mathematics, 8, 1960, p. 181-217.

these planes are linearly independent. It is recalled that

$$P_q = I - A_q (A_q^T A_q)^{-1} A_q^T$$

where I is the unit matrix and A_q the (n,q) matrix whose column vectors are the a^k.

Let x^k be a feasible point lying exactly on the q first boundary planes. If x^k is not a solution of the problem E_2^3 , there are two possibilities (i and ii).

i) $P_q \nabla f(x^k) \neq 0$. This inequality implies $q < n$ (because $P_n = 0$). Then, $P_q \nabla f(x^k)$ is a descent direction and we take (cf. III-1)

$$w^k = - P_q \nabla f(x^k)$$

Then, every point

$$x = x^k + \rho w^k \quad , \quad \rho \geqslant 0$$

is on V_q .

The straight line

$$x = x^k + \rho w^k$$

leaves the feasible region at a point $\bar{x}$ obtained for some ρ, noted $\bar{\rho}$ ($\bar{\rho}$ is strictly positive).

According to (III-1), our purpose is to minimize the function

(20) $$f(x^k + \rho w^k)$$

with respect to ρ, which lies in the interval

$$0 \leqslant \rho \leqslant \bar{\rho}$$

In order to do this, we form

$$w^{kT} \nabla f(x^k + \rho w^k)$$

This product is an increasing function of ρ, because of the convexity of f. For $\rho = 0$, it is

$$w^{kT} \nabla f(x^k) = w^{kT} P_q \nabla f(x^k) = - (w^k)^2 < 0$$

Therefore, on the segment $[x^k, \bar{x}]$, the function (20) has a minimum at a unique point: either $\bar{x}$ or a point between x^k and $\bar{x}$. In these two cases we take this unique point for the $(k+1)^{th}$ approximation, that is x^{k+1}.

ii) $P_q \quad \nabla f(x^k) = 0$, with at least one $y_i < 0$, for instance $y_q < 0$.

We work as in i) by considering the $(q-1)$ first planes, instead of the q first ones.

———————————

OPTIMIZATION THEORY
IN THE DESIGN
OF ELASTIC-PLASTIC STRUCTURES

by

Aleksandras ČYRAS
Vilnius Civil Engineering Institute
USSR

INTRODUCTION

In the mechanics of solids, two general types of problems are dealt with, i.e. those concerning the determination of the stress and strain state of a body subject to various external actions and the problems in which the same values are determined for a body under certain conditions, e.g. plastic failure, locking, etc. The investigation of the latter is referred to as the limit analysis.

The present paper is concerned with the limit state corresponding to plastic failure. It occurs in a body under some loading which will be referred to as the limit load.

The ability of the body to resist any attempt causing this limit state depends on its physical characteristics. For a body of a definite shape such characteristics are the limit forces.

The scalar field of the plasticity constant is considered to be the characteristic of the body as the present investigation is based on the general three dimensional analysis.

The limit analysis implies that, besides the determination of the stress and strain state, the multiplier of the external loads (the body characteristics being prescribed) or the multiplier of the body characteristics (the external loads being prescribed) is to be determined. Both problems are trivial. If the limit load is determined by one parameter too, then it makes no difference which of them is sought.

The first formulation of the problem prevails in structural design in spite of the fact, that the other one could be used more successfully. Two facts seem to account for that. Firstly, the concepts of the limit analysis is formulated on the ground of the observations of the load carrying capacity of various structures; then, a theoretical way of determination of the multiplier of external loads is simple and may be proved experimentally.

Hereinafter, both problems will be referred to as one-parametrical. It is evident that in problems of this kind the value of the multiplier depends on the relationship of the external loads prescribed as well as on the distribution of the plasticity constant of the body. Hence, for a given body it is possible to find a set of combinations of limit external loads, their direction and points of application being prescribed.

By selecting for comparison some optimality criterion, we can evaluate any combination. We are thus led to an extremum problem, the solution of which enables the distribution of the limit external loads corresponding to a certain optimality criterion to be determined. Similarly, the second problem can be dealt with. For a given set of limit external loads, we can find a set of various body compositions (distribution of the plasticity constant) resulting in the limit state. The selection of the optimality criterion corresponding to these combinations will lead to the optimal solution of the problem. It is obvious that the problems are no longer trivial, because they are determined by different optimality criteria.

Now we arrive at the conclusion that the limit analysis includes not only one-parametrical problems, but optimization problems as well. These permit the optimal distribution of the limit external loads or the body characteristics to be determined.

The former will be referred to as the load design problem, the latter as the body design problem.

Note, that a difference should be made between the optimization problems for the limit analysis (and the mechanics of solid deformable body in general) and those for the optimal design of individual structures or their elements. The optimization problems of the analysis are formulated with the same aim in view as one-parametrical problems, i.e. they lead to the determination of the stress and strain states corresponding to the optimal distribution of the external loads or the body characteristics. At the same time the principles of the design method for a particular body are observed, including that all values are considered with respect to the axis or the middle surface.

This is the main difference between a problem of mechanics and that of a conventional optimal design. The latter may be looked upon as the second stage of structural design when the cross sections are designed according to the requirements of building codes and specifications, the optimal (or non-optimal) distribution of forces being taken into account.

Before presenting a detailed formulation of the optimization problems for particular limit states of the material, some peculiarities of the present paper will be discussed.

Most problems in mechanics are formulated by way of the extremum energy principles determining the conditions for the actual stress and strain states. These principles, if expressed by mathematical models, are considered to be extremum problems.

If the problems of this type include some conditions of the form of inequalities, then these should be regarded as being related to mathematical programming.

Certain conditions for the limit design problems are known to be expressed by inequalities, therefore, the analysis may be carried out on the ground of the mathematical programming theory.

The principles of mathematical programming formulated by L.V. Kant rovich in the late thirities and then developed by G.B. Dantzig in the fifties, have recently been used in a number of applied sciences, where the optimal solution of the problems is obtained.

Firstly, the introduction of the method of mathematical programming in some applied sciences makes it possible to formalize and systematize various formulations of problems; secondly, it provides an effective means for a uniform analysis of problems which, when taken separately, required special laborious investigation.

The latter aspect being disregarded, only some statements of the theory of duality concerning mathematical programming are used in the present paper. This theory implies that an extremum problem is compared to the relevant dual problem. A simultaneous consideration of both dual problems proves to be especially useful in the optimal design of extremum problems. It should be noted that an implicit form of the concept of duality is widely used in mechanics.

It is know, for example, that in the mechanics of solids every extremum theorem formulated in terms of force variables has a corresponding dual theorem formulated in terms of strain field.

The conditions for static and kinematic compatibility included in these theorems in some respect are dual, i.e. their operators are conjugate (transposed relative to one another). In mechanics every dual theorem is usually state separately, the calculations being rather complicated. Nevertheless, the use of the general theorems of the theory of duality makes it possible not only to simplify the formulation of the problem, but also to avoid some inexactitude in the formulation itself.

The reader will easily understand the method described in the present paper from a brief investigation of the dual problems of convex programming given in the appendix.

The flow theory of elastic-perfectly plastic bodies is used for the formulation of the theorems and for the derivation of mathematical models (Fig. 1a).

The limit model of the body is that of a rigid-perfectly plastic one (Fig. 1b).

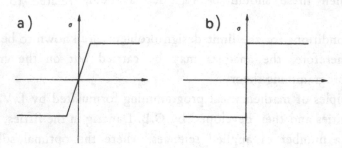

Fig. 1

The optimization problems under investigation being of two different types, it seems quite reasonable to introduce new definitions of primal extremum theorems and consequences resulting from their duality.

The theorems determining the actual stress field will be referred to as static because the conditions include those for static compatibility (equilibrium equations), while the force variables are assumed to be the main unknown values.

The theorems determining the actual strain field will be referred to as kinematic. These include the conditions for kinematic compatibility, the main unknown values being the strains and displacement velocities.

The present paper is mostly concerned with optimization problems, but simultaneously the relevant one-parametrical problems are considered.

The vector-matrix is used. This way of notation is conventional in the mathematic programming theory, and it enables the duality of the problems in question to be shown both for functional and finite- dimensional vector fields. Hence, the tensor of stresses, strains and displacements dealt with in the mechanics of continua may be presented in the form of vector functions, every component of which is assumed to be function of the coordinates of a given point.

The notation adopted simplifies the statements and provides a convenient means for the computation of the finite-dimensional models of the problems by the method of finite differences, finite elements, nets of failure, etc.

I GENERAL STATEMENTS

1.1. Basic Definitions

Consider an isotropic solid of the volume V with respect to the Cartesian coordinates $x \equiv (x_1, x_2, x_3)$. The surface S of the body is divided into two parts: S_p which is subjected to loading, S_u where the displacements are prescribed completely. Hereinafter, the displacements on S_u are assumed to be zero.

Monotonically increasing loading is a set of forces, each proportional to the same parameter and increasing from zero to a certain value. Under monotonically increasing loading the set of forces will be defined by the vector field $P \equiv (P_1, P_2, P_3)^T$ which may be prescribed either completely or only to within the scalar multiplier P, its direction $e \equiv (e_1, e_2, e_3)^T$ being prescribed. The condition $P = Pe$ holds for the latter case.

In case of monotonically increasing loading, the limit state of a plastic body is reached due to the plastic strains increasing infinitely, the stress field being constant, i.e. simple plastic failure occurs.

Cyclic loading is a set of forces which, if taken separately or in groups, may be varied independently within prescribed limits.

The upper bound of the forces is defined by the vector field $P^+ \equiv (P_1^+, P_2^+, P_3^+)^T$ the lower one by the vector field $P^- \equiv (P_1^-, P_2^-, P_3^-)^T$.

In case of cyclic loading, every set of forces does not necessarily lead to plastic failure, nevertheless, certain combinations may cause cyclic plastic deformation resulting in plastic failure. Thus, the body reaches its limit state at cyclic plastic failure. This definition applies to the so called "progressive failure", when the sign of the deformation remains unchanged, the deformation increasing in every cycle, as well as to "sign-variable plasticity", in which case the deformations have alternative signs.

Now it should be noted, that the upper and lower bounds of the loadings may be prescribed to within a certain scalar multiplier, their direction $e \equiv (e_1, e_2, e_3)^T$ being prescribed. The condition $\underline{P}^+ = P^+ e$ holds for the latter case. Keeping in mind the above condition, the displacements on S_u being zero, we shall present the external reactions in the form of the vector field of response, which may be either actual $R \equiv (R_1, R_2, R_3)^T$ or residual $r \equiv (r_1, r_2, r_3)^T$.

The stress field of the body is defined by the vector field of the actual stresses $\sigma \equiv (\sigma_{11}, \sigma_{22}, \sigma_{33}, \sigma_{12}, \sigma_{23}, \sigma_{31})^T$.

If the body is assumed to be in an elastic state, then the extreme values of the stresses will be expressed as follows - the upper bound $\vec{\sigma} \equiv (\sigma_{11}^+, \sigma_{22}^+, \sigma_{33}^+, \sigma_{12}^+, \sigma_{23}^+, \sigma_{31}^+)^T$ and the lower bound $\vec{\sigma} \equiv (\sigma_{11}^-, \sigma_{22}^-, \sigma_{33}^-, \sigma_{12}^-, \sigma_{23}^-, \sigma_{31}^-)^T$. The residual stresses produced in the body after it has reached its limit state and when it is no longer subject to external actions, are defined by the vector field

$$\rho \equiv (\rho_{11}, \rho_{22}, \rho_{33}, \rho_{12}, \rho_{23}, \rho_{31})^T.$$

The strain field of the body is defined by the vector field of actual strains $\epsilon \equiv (\epsilon_{11}, \epsilon_{22}, \epsilon_{33}, \epsilon_{12}, \epsilon_{23}, \epsilon_{31})^T$ and displacements $u \equiv (u_1, u_2, u_3)^T$. If the body is assumed to be in an elastic state, then, the extremum values of the strains will be expressed as follows: .the upper bound $\vec{\epsilon}^+ \equiv (\epsilon_{11}^+, \epsilon_{22}^+, \epsilon_{33}^+, \epsilon_{12}^+, \epsilon_{23}^+, \epsilon_{31}^+)^T$ and the lower bound $\vec{\epsilon}^- \equiv (\epsilon_{11}^-, \epsilon_{22}^-, \epsilon_{33}^-, \epsilon_{12}^-, \epsilon_{23}^-, \epsilon_{31}^-)^T$. The vector fields of the residual strains and residual displacements are expressed by $\xi \equiv (\xi_{11}, \xi_{22}, \xi_{33}, \xi_{12}, \xi_{23}, \xi_{31})^T$ and $w \equiv (w_1, w_2, w_3)^T$ respectively.

Every component of the above values is a function of the coordinates of a point of the body, e.g. $\sigma_{11} \equiv \sigma_{11}(x), \epsilon_{11} \equiv \epsilon_{11}(x), u_1 \equiv u_1(x)$ etc. Besides, some of these values may depend on time too, in this case the argument t will be introduced, e.g. $\sigma(x,t), \epsilon(x,t), u(x,t)$ etc. The velocities (increments per unit of time) of the corresponding values are denoted by a dot, $\dot{\sigma}, \dot{\epsilon}, \dot{u}, \dot{P}$ etc.

It has already been mentioned, that the characteristic of a plastic body is the plasticity constant C. This is the measure of the yield point of the material, and it may be regarded as a scalar field determined in the volume V, i.e. it is $C \equiv C(x)$.

The definition of the stress and strain field in the volume V leads to the energy value called energy dissipation. This is the dissipation energy at plastic deformation expressed by the scalar product of the stress and strain fields, its value being essentially positive:

(1.1)
$$D = \int_V \epsilon^T \cdot \sigma \, dv.$$

The fields of external loading and displacements on the surface S_p determine the energy value called the work of external actions:

(1.2
$$W = \int_{S_p} P^T \cdot u \, ds.$$

Taking into consideration that the directions P and U are assumed to have appropriate signs, we may conclude that the work of external actions is an essentially positive value.

If one of the fields in the equations (1.1) and (1.2) stands for the velocities of the corresponding values, then the former may be referred to as plastic dissipation (dissipation power) and the latter - as the power of external actions (loads or displacements).

1.2. Basic Assumptions and Relationships

The deformations of the body are assumed to be small, hence the changes in the geometry, when deriving the equilibrium equations, may be disregarded. Let us determine the differential operator of static compatibility as the matrix:

$$
\nabla \equiv \left\|
\begin{array}{cccccc}
\partial/\partial x_1 & 0 & 0 & \partial/\partial x_2 & 0 & \partial/\partial x_3 \\
0 & \partial/\partial x_2 & 0 & \partial/\partial x_1 & \partial/\partial x_3 & 0 \\
0 & 0 & \partial/\partial x_3 & 0 & \partial/\partial x_2 & \partial/\partial x_1
\end{array}
\right\|
\tag{1.3}
$$

Then the static compatibility of the stresses in the volume V will be expressed by the linear differential equation, the volume forces being disregarded:

$$
\nabla \sigma = 0.
\tag{1.4}
$$

The kinematic compatibility of the deformations and displacements in the volume takes the form:

$$
\epsilon = \nabla^T u,
\tag{1.5}
$$

where ∇^T is the transposed operator.

Let us determine the algebraic operator of the static compatibility on the surface S of the body as the matrix:

$$
N \equiv \left\|
\begin{array}{cccccc}
n_1 & 0 & 0 & n_2 & 0 & n_3 \\
0 & n_2 & 0 & n_1 & n_3 & 0 \\
0 & 0 & n_3 & 0 & n_2 & n_1
\end{array}
\right\|
\tag{1.6}
$$

where $n \equiv (n_1, n_2, n_3)$ is the unit vector field of the external normal to the surface S. Hence the boundary conditions on the surface S_p for the load and those on S_u for the reactions will be:

(1.7)
$$N\sigma = P,$$
$$N\sigma = R.$$

Note that the structure of the matrices (1.3) and (1.6) is similar.

If external actions depend on time, their velocities being sufficiently low, then the problem may be looked upon as being quasistatic and the equations (1.4) and (1.5) will be valid.

The consideration of cyclic external actions is based on the values of the extremum stress fields, when the behaviour of the body is elastic. The solution of the elastic boundary problem is assumed to be known. Let us denote the influence matrix of the elastic solution by $\omega \equiv \omega(x)$; then the stress fields will be:

(1.8)
$$\sigma = \int_{S_p} \omega P \, ds$$

Making use of the extremum values of the load P^+ and P^- which are independent of time, we obtain the vectors σ^+ and σ^- determining the extremum values of "elastic" stresses:

(1.9)
$$\left. \begin{array}{l} \sigma^+ = \int_{S_p} (\omega^+ P^+ - \omega^- P^-) \, ds, \\[2mm] \sigma^- = \int_{S_p} (\omega^- P^+ - \omega^+ P^-) \, ds, \end{array} \right\}$$

where ω^+ and ω^- are influence operators of the "elastic" design obtained from various combinations of the vectors ω_j of the operator ω.

It follows that:

$$\omega_j^- = 0 \quad \text{subject to} \quad \omega_j^+ = \omega_j$$

and $\qquad \omega_j^+ = 0 \qquad$ subject to $\qquad \omega_j^- = \omega_j$

Thus, the equations (1.9) determine the surface of the extremum "elastic" stresses which is usually symmetric with respect to its centre.

The equation (1.9) implies that the lower bound of the load whose direction is actually negative, is positive too, i.e. $\mathbf{P}^- \equiv -\mathbf{P}^- \geqslant 0$. This change of the sign is convenient when the problems are formulated as dual pairs of mathematical programming problems.

Then the stress values obtained through (1.9) will have their actual sign. The behaviour of the material corresponding to the complex stress field is expressed by a function the arguments of which are these stresses. The relationship between this function and the constant of the material (the body characteristic) is expressed by the yield conditions:

$$f(\sigma) \leqslant C, \tag{1.10}$$

The function f is assumed to be concave. Therefore, the set (domain) expressed by the inequality (1.10) is convex. If the expression (1.10) is understood as an equality, then the yield condition in the stress field will be represented by a certain surface. If the function f is convex, this surface will be flat, and every point of it corresponds to one gradient value only. In case the function is piece-wise regular smooth, there may be singular irregular surfaces corresponding to different gradient values.

The scalar multiplier is used here as a scale, the value D_i may be written in the form of the scalar product:

$$\dot{D}_i = \sigma^T \cdot \lambda \frac{\partial f(\sigma)}{\partial \sigma} \tag{1.11}$$

Then, denoting

$$\lambda \frac{\partial f(\sigma)}{\partial \sigma} = \dot{\epsilon} \tag{1.12}$$

we can see that the expression (1.11) determines the plastic dissipation at the point i of the elementary unit volume of the body. Now it is to be proved, that from a physical point of view ϵ in (1.12) expresses the vector field of the strain rates. If this statement is true (it will be proved below), then we may conclude that a unique stress field corresponds to more than one strain rate fields when the function is piecewise regular smooth.

The plastic dissipation in the volume V may be presented in a different way by:

$$(1.13) \qquad\qquad \dot{D} = \int\limits_{V} \lambda C$$

Eq. (1.13) follows from the following relations on the yield surface which hold for a constant multiplier:

$$(1.14) \qquad\qquad \sigma^{T} \cdot \dot{\epsilon} = \sigma^{T} \cdot \lambda \, \frac{\partial f(\sigma)}{\partial \sigma} = \lambda f(\sigma) = \lambda C$$

As the value C and the multiplier used as a scale are positive, the expression (1.14) shows once more that the plastic dissipation is an essentially positive value.

Now some definitions concerning the stress, strain and displacement fields will be presented.

The stress field (or the stress rates) is statically possible, if it satisfies the differential equilibrium equation (1.4) and the primal boundary conditions.

The stress field (or the stress rates) is admissible if it satisfies the constraints as inequalities. The constraints may be, for example, the yield conditions of a plastic body, or of any other type on the boundary S_p.

The field is statistically admissible, if it satisfies the conditions both for statically admissible and statically possible fields taken together.

The field of displacements (strains) and their rates is kinematically possible, if it satisfies the kinematic compatibility conditions (1.5) and the primal boundary conditions on the surface.

The field of displacements (strains) and their rates is admissible, if it satisfies the constrains as inequalities.

The field of displacements (strains) is kinematically admissible, if it satisfies the conditions both for statically possible and statically admissible fields taken together.

Now it will be noted, that in contrast to the elastic theory, in the formulation of problems for plastic bodies there may occur discontinuities of stresses, strains and displacements as well as of their rates.

In this case the plastic dissipation in the volume must be added to the dissipation on the discontinuity surface. This applies to particular cases only, and hereinafter, in the problem where the discontinuity of surfaces may occur, the expression (1.1) for the energy dissipation will be understood as including the dissipation on the discontinuity surface.

1.3. Optimality Criterion

Let us formulate the analytical expressions for the optimality criterion in the load design and body design problems of the limit analysis.

In the load design problem the value to be varied is the loading on the surface S_p. We shall assign to each load a certain weight multiplier measuring the influence of a unit value of this load in the optimal load distribution.

This value will be referred to as the weight multiplier of the optimality criterion.

The external action at a point on the surface S_p being determined by a vectorial quantity, hence the weight multiplier for this point must be presented as a vector. Therefore, the optimal distribution of the external action is defined by the vector of the weight multipliers.

Let this field be $T \equiv (T_1, T_2, T_3)^T$ under monotonically increasing loading. The optimality criterion of the load design problem consists of maximizing the scalar product of the vector fields of the weight multipliers by the load.

$$\max \int_{S_p} T^T \cdot P \, eds, \qquad (1.15)$$

The expression (1.15) determines the optimal distribution of the limit load, i.e. of the scalar field P, its direction being prescribed.

If the loading is cyclic, the value of the weight multiplier for the upper and lower bounds may be different. Let the vector fields be $T^+ \equiv (T_1^+, T_2^+, T_3^+)^T$ and $T^- \equiv (T_1^-, T_2^-, T_3^-)^T$ respectively. The optimality criterion for an elastic-plastic body has the form:

$$\max \left\{ \int_{S_p} T^{+T} \cdot P^+ \, eds + \int_{S_p} T^{-T} \cdot P^- eds \right\} \qquad (1.16)$$

Note that the above optimality criteria enable different optimization problems for the distribution of the limit load to be obtained, depending on the selection of various fields of the weight multipliers. Let, for example, the vector field of the weight multipliers be unity throughout the surface S_p of a plastic body. If we denote it by T_e then $|T_e| \equiv 1$ at every point of the surface S_p as the condition implies. The optimality criterion:

$$\max \int_{S_p} T_e^T \cdot Pe\,ds$$

may be used in the derivation of the problem determining the maximum load to be applied to the surface S . Nevertheless, this is only an individual case of the load design problem where the optimality criterion is in general expressed by (1.15).

In the body design problem the value to be varied is the constant of the material. The volume V of the body is assumed to be constant. It is necessary to find the optimality criterion determining the quality of the body, corresponding to various distributions of its constant. For every point of the body, we shall choose the weight multiplier related to the optimal value of the constant at that given point. Because C(x) is a scalar field, the weight multipliers in the entire body will also form a scalar field. Let $\Lambda \equiv \Lambda(x)$ be the scalar field of the weight multipliers of the optimality criterion of the body. The optimality criterion for the body design problem consists of minimizing the product of the fields of the weight multipliers by the body constant. Hence the optimality criterion for a plastic body has the form:

(1.17) $$\min \int_V \Lambda C\,dv,$$

The expression (1.17) is the analogue of the optimal distribution of the constant of the material in the volume prescribed. It defines the best composition of the body, the minimum values of its stress and strain characteristics being satisfied. If by proper selection of the weight multipliers these characteristics are related to some extent to the cost of the material of the body, then the optimality criterion (1.17) expresses the minimum cost of the body, its volume being fixed.

It should be noted that an optimization problem of a solid may be understood as that of a minimum volume (weight). For a plastic-rigid body, it may be stated as follows: it is necessary to find the configuration of a free part of the surface of a body, that the volume be minimum and C be constant at the limit state.

The "minimum cost" design problems (V is constant, C is unknown) and "minimum volume" design problems (C is constant, V is to be determined) for framed systems, plates and shells are understood as being the alternatives of the general optimization design problem for the solid satisfying the criterion (1.17). This results from the fact that in such problems all the values are considered with respect to the axis or to the middle surface of the structure. The design problems for such structures, therefore lead to the derivation of the optimal distribution of the limit forces, corresponding to the limit state, the external actions being prescribed.

Nevertheless, the limit forces corresponding to a certain cross-section are functions of a specific dimension (e.g. the height, the beam cross-section or the plate thickness h) and a body characteristic (e.g. yield point σ_T) i.e. $Q = Q(h, \sigma_T)$. Assuming that value of h is prescribed (h being constant), we can probably select the values of the weight multipliers in order that the minimum cost of the constitutive material will be expressed by the optimality criterion (1.17). If we assume the material of the structure to be homogeneous (σ_T being constant), then on the ground of the relationship between the volume and the dimension h we can take the values of Λ leading to the minimum volume of the structure. This applies only to particular problems which may be successfully solved by way of the optimality criterion (1.17) when "the minimum cost" or "the minimum volume" are sought.

II AN ELASTIC-PLASTIC BODY UNDER MONOTONICALLY INCREASING LOADING

The mathematical models of the optimization problems for a rigid-plastic body under monotonically increasing loading will be derived. The limit state of the body is reached when plastic failure occurs. The stress and strain fields of the body at plastic failure are defined by the vector fields of the stresses σ, the displacement velocities $\dot{u}$ and the strain rates $\dot{\epsilon}$. These are the main values to be found. Besides, the field of the load P on the surface S_p (the load direction being prescribed) is sought in the load design problem; the scalar field of the plasticity constant C is the value to be found in the body design problem. The unknown values are written on the lefthand side of the relations, the prescribed ones on the right-hand side.

2.1. Load Design Problem
2.1.1. Static Formulation of the Problem
To derive the mathematical model, we shall use the extremum energy principle called the static theorem of the limit load.

Of all statically admissible stress fields at simple plastic failure, that one is actual which corresponds to the maximum value of external power.

The theorem may be written in the form of the extremum problem:

(2.1)
$$\max \int_{S_p} \dot{u}^T \cdot P ds$$

subject to

(2.2)
$$\left. \begin{array}{l} f(\sigma) \leqslant C \\ \nabla\sigma = 0 \end{array} \right\} \text{ in } V,$$

$$\left. \begin{array}{l} N\sigma - R = 0 \quad \text{on } S_u \; ; \\ N\sigma - P = 0 \quad \text{on } S_p \; . \end{array} \right\}$$

Now the relations included in (2.2) will be discussed. The first one, i.e. $f(\sigma) \leqslant C$

defines the admissible stress field, the remaining refer to the statically possible field. Thus the conditions (2.2) as a whole determine the statically admissible stress field. There are many fields satisfying the conditions (2.2), and as the above theorem implies, the actual one maximises the value of the external power, i.e. expression (2.11). Now let us turn to the derivation of the mathematical model for the load design problem on the ground of the extremum problem (2.1)-(2.2). The displacement velocity is not included in the conditions (2.2), the maximization of the expression (2.1) is, therefore, possible for any fixed values of u, including $u \equiv T$ where T is the vector field of the weight multipliers of the optimality criterion of the load.

Assuming that the direction of the load on S_p is prescribed, i.e. $P = Pe$, $P \geqslant 0$ we obtain:

$$\max \int_{S_p} T^T \cdot Pe \, ds \qquad\qquad (2.3)$$

subject to

$$\left. \begin{array}{c} f(\sigma) \leqslant C \\ \nabla\sigma = 0 \end{array} \right| \text{in V,} \\ N\sigma - R = 0 \quad \text{on } S_u, \\ \left. \begin{array}{c} N\sigma - Pe = 0 \\ P \geqslant 0 \end{array} \right| \text{on } S_p, \qquad (2.4)$$

This is the mathematical model of the load design problem for a rigid-plastic body under monotonically increasing loading. The problem itself is the functional analogue of a convex programming problem, where the linear functional (2.3) is maximized on the convex set (2.4). The solution of the problem enables the scalar field of the limit load values as well as the vector field of the actual stresses corresponding to the state of plastic failure to be determined. Hereinafter, the problem (2.3)-(2.4) will be referred to as the static formulation of the problem.

2.1.2. Kinematic Formulation

The kinematic formulation of the problem (2.3)-(2.4) may be obtained iormally on the ground of the theory of duality for functional analogues of convex programming problems.

Let us derive the Lagrange functional for the problem (2.3)-(2.4). With this end in view new variables, i.e. the multipliers $\lambda, \beta, \gamma, \mu, \nu$ will be introduced. The functional consists of the expression (2.3) and the conditions (2.4) multiplied by the corresponding multipliers. Thus we arrive at:

$$F(P,\sigma,R,\lambda,\beta,\gamma,\mu,\nu) = \int_{S_p} T^T \cdot Pe\,ds +$$

(2.5)
$$+ \int_V \lambda[C - f(\sigma)]\,dv + \int_V \beta^T \cdot \nabla\sigma\,dv + \int_{S_u} \gamma^T \cdot (N\sigma - R)\,ds +$$

$$+ \int_{S_p} \mu^T (N\sigma - Pe)\,ds + \int_{S_p} \nu \cdot P\,ds.$$

Here, the Lagrange multipliers are expressed by:

(2.6)
$$\left.\begin{array}{l}
\left.\begin{array}{l}
\lambda \geqslant 0 \\[6pt]
\beta \equiv (\beta_1,\beta_2,\beta_3)^T \gtreqless 0
\end{array}\right\} \quad \text{in } V, \\[14pt]
\gamma \equiv (\gamma_1,\gamma_2,\gamma_3)^T \gtreqless 0 \qquad \text{on } S_u, \\[10pt]
\left.\begin{array}{l}
\mu \equiv (\mu_1,\mu_2,\mu_3)^T \gtreqless 0 \\[6pt]
\nu \geqslant 0
\end{array}\right\} \quad \text{on } S_p,
\end{array}\right\}$$

Their physical sense can be easily shown. As the first term $\int_{S_p} T^T \cdot Pe\,ds$ of the Lagrange functional (2.5) is an energy value and a scalar, the dimension of other terms must be the same. It follows that the expressions $\lambda \equiv \lambda(x)$ and $\nu \equiv \nu(x)$ define the displacement velocities, i.e. the vector fields. Then we have:

(2.7)
$$\dot{u} \equiv \begin{cases} \beta & \text{in } V, \\ \gamma & \text{on } S_u, \\ \mu & \text{on } S_p. \end{cases}$$

Hence the Lagrange functional takes the form:

$$F(P, \sigma, R, \lambda, \dot{u}, \nu) = \int\limits_{S_p} T^T \cdot Pe\, ds +$$

$$+ \int\limits_V \lambda\, [\, C - f(\sigma)\,]\, dv + \int\limits_V \dot{u}^T \cdot \nabla \sigma\, dv + \qquad\qquad (2.8)$$

$$+ \int\limits_{S_u} \dot{u}^T \cdot (N\sigma - R)\, ds + \int\limits_{S_p} \dot{u}^T \cdot (N\sigma - Pe)\, ds + \int\limits_{S_p} \nu \cdot P\, ds\,.$$

The problem dual to (2.3) - (2.4) may be formulated as follows: it is necessary to find

$$\min_{(\lambda, \mu)} F_1$$

subject to:

$$\left.\begin{aligned} F_1 &= \max_{(P, \sigma, R)} F \\ \lambda &\geqslant 0 \ \text{in V,} \\ \nu &\geqslant 0 \ \text{on } S_p\,, \end{aligned}\right\} \qquad\qquad (2.9)$$

The conditions of the dual problem are, therefore, the equilibrium conditions (2.5) determining the arbitrary variations of the load on the surface S_p the stresses in the volume V and the response on the surface S_u. It is known that the necessary and sufficient condition for the functional to be stationary is that its variations with respect to the corresponding variables must be zero.

As the variations of δP on S_p of $\delta \sigma$ in V and of δR on S_u are arbitrary, we have:

$$e^T \cdot T - e^T \cdot \dot{u} + \nu = 0, \quad \nu \geqslant 0 \text{ on } S_p ,$$

(2.10)
$$- \lambda \frac{\partial f(\sigma)}{\partial \sigma} + \nabla^T \dot{u} = 0, \quad \lambda \geqslant 0 \text{ in } V,$$

$$\dot{u} = 0 \qquad \text{on } S_u .$$

Here the variation of the expression $\int\limits_V \dot{u}^T \cdot \nabla\sigma \, dv$ with respect to σ is carried out taking account of:

(2.11)
$$\int\limits_V \dot{u}^T \cdot \nabla\sigma \, dv = \int\limits_V \sigma^T \cdot \nabla^T \dot{u} \, dv \pm \int\limits_S \sigma^T \cdot N^T \dot{u} .$$

Thus, substituting the expressions (2.10) in (2.8) we obtain:

(2.12)
$$\min \left\{ \int\limits_V \left[\lambda \frac{\partial f(\sigma)}{\partial \sigma} \right]^T \cdot \sigma \, dv + \int\limits_V \lambda [C - f(\sigma)] \, dv \right\}$$

subject to:

(2.13)
$$\left. \begin{array}{c} \nabla^T \dot{u} - \lambda \dfrac{\partial f(\sigma)}{\partial \sigma} = 0 \\ \lambda \geqslant 0 \end{array} \right\} \text{ in } V,$$

$$\dot{u} = 0 \quad \text{on } S_u ,$$

$$\dot{u} \geqslant T \quad \text{on } S_p .$$

This is the kinematic formulation of the mathematical model of the load design problem, which is dual to the problem (2.3) - (2.4). The vector fields of the stress σ, the displacement velocities $\dot{u}$ and the scalar field of the multipliers λ are sought. Now the physical sense of the problem will be explained.

The first and the third relations in the conditions (2.13) define the kinematically possible field, the second and the fourth relations stand for the admissible field of the displacement velocities. This becomes evident if in the first condition (2.13) we denote:

$$\lambda \frac{\partial f(\sigma)}{\partial \sigma} = \dot{\epsilon}$$

Thus, all conditions taken together determine the kinematically admissible field.

The expression (2.12) may be understood as defining the plastic dissipation, because the first term is:

$$\int_V \left[\lambda \frac{\partial f(\sigma)}{\partial \sigma} \right]^T \cdot \sigma = \int_V \dot{\epsilon}^T \cdot \sigma \mathrm{d}v = \dot{D}, \qquad (2.14)$$

the second term for the optimal solution is zero on the ground of the second theorem of duality. Hence, the extremum problem (2.12) - (2.13) corresponds to the following extremum theorem:

Of all kinematically admissible fields of displacement velocities at simple plastic failure, that one is actual which corresponds to the minimum value of plastic dissipation.

This principle will be referred to as the kinematic theorem of limit load.

1. Theorems of Duality

The mathematical models (2.3) - (2.4) and (2.12) - (2.13) being dual, the static and kinematic theorems of limit load are dual too.

Now the theorems dual to these and to the corresponding mathematical models will be stated.

According to the second theorem of duality (additional orthogonality condition), the following condition must be satisfied for the optimal solution of λ^* and σ^*

$$\lambda^*[C - f(\sigma^*)] = 0 \qquad (2.15)$$

This condition is similar to those of the associated plastic flow rule:

$$(2.16) \qquad \left. \begin{array}{l} \lambda > 0 \quad \text{subject to} \quad f(\sigma) = C, \\ \lambda = 0 \quad \text{subject to} \quad f(\sigma) < C. \end{array} \right\}$$

These relations will be referred to as the second theorem of duality for a rigid-plastic body.

According to the first theorem of duality, the values of the objective function for the optimal solutions of dual problems are equal, i.e.

$$(2.17) \qquad \int_{S_p} T^T \cdot P^* e \, ds = \int_V \left[\lambda^* \frac{\partial f(\sigma^*)}{\partial \sigma} \right]^T \cdot \sigma^* \, dv.$$

The expression (2.17) shows that the external power is equal to the plastic dissipation. This will be referred to as the first theorem of duality of the load design problem for rigid-plastic body.

Now the generalized Lagrange problem will be written for the static and kinematic formulations of the dual pair of problems obtained. It consists of all the conditions for both problems and the additional orthogonality condition, i.e. the conditions (2.16):

$$(2.18) \qquad \left. \begin{array}{l} \left. \begin{array}{l} f(\sigma) \leqslant C \\ \nabla \sigma = 0 \\ \nabla^T \dot{u} - \lambda \dfrac{\partial f(\sigma)}{\partial \sigma} = 0 \\ \lambda \geqslant 0 \\ \lambda > 0, \text{ subject to } f(\sigma) = C \\ \lambda = 0, \text{ subject to } f(\sigma) < C \end{array} \right\} \text{ in } V, \\[2em] \left. \begin{array}{l} N\sigma - R = 0 \\ \dot{u} = 0 \end{array} \right\} \text{ on } S_u, \\[1.5em] \left. \begin{array}{l} \dot{u} \geqslant T \\ N\sigma - Pe = 0 \\ P \geqslant 0 \end{array} \right\} \text{ on } S_p, \end{array} \right\}$$

The fields $P, \sigma, \dot{u}$ and λ satisfying these relations provide the solution of the load design problem for a rigid-plastic body. It should be noted that the explicit form of the physical principle (2.14) is not included in (2.18), and this means that it is not needed for the solution of the problem. If an extremum theorem is formulated for the prescribed yield conditions, then the physical relationship between the strain rates and the stresses (associated flow rule) is the consequence of the dual relations of this theorem. The validity of this theorem must, of course, be proved.

2.1.4. Proof of Static Theorem of Limit Load

The proof of the static theorem is exceptionally simple. The mathematical models of the load design problem based on the Kuhn-Tucker relations are derived through the static theorem.

Let us assume that the actual field satisfying the conditions (2.4) does not correspond to the maximum value of the expression (2.36). The solution of the extremum problem (2.36) - (2.4) σ^* is denoted by an asterisk. The fields σ and σ^* are assumed to correspond to a single field of the limit load. Then according to the Kuhn-Tucker theorem, the Lagrange functional must satisfy the inequality:

$$F(P^*, \sigma^*, R^*, \lambda^*, \dot{u}^*, \nu^*) - F(P, \sigma, R, \lambda^*, \dot{u}^*, \nu^*) \geqslant 0. \qquad (2.19)$$

Substituting the expression (2.8) in (2.19) and bearing in mind that P is equal to P^* we have:

$$\int_V \lambda^* [f(\sigma) - f(\sigma^*)] dv \geqslant 0. \qquad (2.20)$$

Since the optimal solution of the problem (2.3) - (2.4) satisfies the condition $\lambda^* f(\sigma^*) = \lambda^* C$, the inequality (2.20) yields:

$$f(\sigma) \geqslant C.$$

This contradicts the yield conditions. Thus the expression (2.20) may be satisfied only as an equality:

$$f(\sigma) = f(\sigma^*).$$

Taking into consideration that σ and σ^* satisfy the same linear equations of equilibrium, we obtain:

$$\sigma = \sigma^*. \qquad (2.21)$$

Hence, the actual field at simple plastic failure maximizes the value of the external power. The theorem has been proved.

2.1.5. One-Parametrical Problem

The above mathematical models of the load design problem are derived through the limit load theorem.

This theorem may be used for the derivation of the models for one-parametrical problems. Let the load distribution on the surface S_p be expressed by

$$(2.22) \qquad P = P^\circ \eta.$$

where P_o is the required parameter and η is the prescribed vector field of the load distribution. Thus, the value of the limit load depends only on the parameter P. The expression (1.2) defining the external power takes the form:

$$(2.23) \qquad W = P^\circ \int_{S_p} \dot{u}^T \cdot \eta \, ds.$$

Since the variable $\dot{u}$ is not included in (2.2), then fixing the value of the multiplier $\int_{S_p} \dot{u}^T \cdot \eta \, ds \equiv 1$, we obtain the dual pair of problems determining the limit load multipliers:

a) **Static Formulation**

$$(2.24) \qquad \max P^\circ$$

subject to

$$(2.25) \qquad \begin{aligned} f(\sigma) &\leqslant C \\ \nabla \sigma &= 0 \end{aligned} \Bigg\} \ \text{in V,} \\ N\sigma - R = 0 \quad \text{on } S_u, \\ \begin{aligned} N\sigma - P^\circ \eta &= 0 \\ P^\circ &\geqslant 0 \end{aligned} \Bigg\} \ \text{on } S_p,$$

b) Kinematic Formulation

$$\min \left\{ \cdot \int_V \left[\lambda \, \frac{\partial f(\sigma)}{\partial \sigma} \right]^T \cdot \sigma \, dv + \int_V \lambda \left[C - f(\sigma) \right] dv \right\} \tag{2.26}$$

subject to:

$$\left. \begin{array}{r} \nabla^T \dot{u} - \lambda \, \dfrac{\partial f(\sigma)}{\partial \sigma} = 0 \\[2mm] \lambda \geqslant 0 \end{array} \right\} \text{ in } V, \\[2mm] \left. \begin{array}{r} \dot{u} = 0 \quad \text{ on } S_u, \\[2mm] \int_{S_p} \dot{u}^T \cdot \eta \, ds \geqslant 1 \quad \text{ on } S_p. \end{array} \right\} \tag{2.27}$$

Thus, the mathematical models of a one-parametrical load design problem has been obtained. The principal theorems of duality may be applied to this dual pair of problems. The first one

$$P^{o*} = \int_V \left[\lambda^* \, \frac{\partial f(\sigma^*)}{\partial \sigma} \right]^T \cdot \sigma^* \, dv \tag{2.28}$$

implies that the limit load parameter is equal to the plastic dissipation for the actual field of stress and strain rates, the field of the displacement velocities on S_p being fixed. If for fixing the expression $\int_{S_p} \dot{u}^T \cdot \eta \, ds$ we choose any value a, rather than unity, then the relation $\int_{S_p} u^T \cdot \eta \, ds \geqslant a$ may be considered as the fixed displacement velocity on S_p and the relation $P^o \int_{S_p} \dot{u}^T \cdot \eta \, ds \geqslant P^o a$ as the fixed power. The former seems to be more reasonable.

The generalized Lagrange problem for the dual pair of one-parametrical problem has the form:

$$\left.\begin{array}{r}
f(\sigma) \leqslant c \\[4pt]
\nabla\sigma = 0 \\[4pt]
\nabla^T \dot{u} - \lambda \dfrac{\partial f(\sigma)}{\partial \sigma} = 0 \\[4pt]
\lambda \geqslant 0 \\[4pt]
\lambda > 0, \text{ subject to } f(\sigma) = c \\[4pt]
\lambda = 0, \text{ subject to } f(\sigma) < c
\end{array}\right\} \text{ in } V,$$

(2.29)

$$\left.\begin{array}{r}
N\sigma - R = 0 \\[4pt]
\dot{u} = 0
\end{array}\right\} \text{ on } S_u,$$

$$\left.\begin{array}{r}
\displaystyle\int_{S_p} \dot{u}^T \cdot \eta\, ds \geqslant 1 \\[4pt]
N\sigma - P^\circ \eta = 0 \\[4pt]
P^\circ \geqslant 0
\end{array}\right\} \text{ on } S_p,$$

2.1.6. Comparison of Conditions of Optimization and One-Parametrical Problems

A comparison between the one-parametrical load-design problem (2.23) and the optimization problem (2.18) shows that the conditions of both problems are similar in the volume V and on the surface S_p the difference being only on the surface S_u. In the optimization problem, the constraints on the displacement velocities on S_p are expressed by the vector inequality $\dot{u} \geqslant T$, in the one-parametrical problem - by the integral expression $\int_{S_p} \dot{u}^T \cdot \eta\, ds \geqslant 1$ Both constrains enable the field of the displacement velocities to be determined to within a constant multiplier.

Moreover, the constraint $\dot{u} \geqslant T$ shows that the weight multipliers of the optimality criterion are obtained from the kinematic formulation and express the lower bound for every component of the displacement velocities on S_p. In the one-parametrical problem this deformation constraint is expressed by the reduced displacement velocity throughout the surface (in the case under consideration the velocity is equal to unity). It is obvious, that the first constraint is tighter with respect to the variables u, but at the same time it makes possible a free selection of the constraints and thus permits an optimal distribution of the limit load to be obtained.

2.2. Body Design Problem

2.2.1. Static Formulation

To derive the mathematical model, we shall use the extremum energy principle referred to as the static theorem of simple plastic failure.

Of all statically admissible stress fields, that one is actual at simple plastic failure, which corresponds to the minimum value of plastic dissipation.

The theorem may be written in the form of the extremum problems

$$\min \int_V \lambda C \, dv \tag{2.30}$$

subject to

$$
\left.
\begin{aligned}
C - f(\sigma) &\geq 0 \\
C &\geq 0 \\
\nabla \sigma &= 0
\end{aligned}
\right\} \text{ in } V, \\
\left.
\begin{aligned}
N\sigma - R &= 0 \quad \text{on } S_u, \\
N\sigma &= P \quad \text{on } S_p .
\end{aligned}
\right\}
\tag{2.31}
$$

In the body design problem, the field of the loading on S_p is assumed to be prescribed completely. The scalar field of the plasticity constant in V is sought. The extremum problem (2.30) - (2.31) is used to derive the mathematical model determining the optimization problems. The procedure is similar to that in the load design problem. Since the scalar field of the multipliers λ is not included in the conditions (2.31), then the minimization of (2.30) is possible at any fixed value of them, including $\lambda \equiv \Lambda$, where Λ is the scalar field of the weight multipliers of the optimality criterion of the body.

Thus, we have:

$$\min \int_V \Lambda C \, dv \tag{2.32}$$

subject to

$$
\left.
\begin{array}{r}
C - f(\sigma) \geqslant 0 \\
C \geqslant 0 \\
\nabla\sigma = 0
\end{array}
\right\} \text{ in } V,
$$

(2.33)

$$
\left.
\begin{array}{r}
N\sigma - R = 0 \quad \text{on } S_u, \\
N\sigma = P \quad \text{on } S_p.
\end{array}
\right\}
$$

This is the mathematical model of the body design problem for a rigid-plastic body under monotonically increasing loading. The problem is the functional analogue of a convex programming problem. Here, the linear functional (2.32) is minimized the convex set (2.33). The problem will be referred to as the static formulation of the body design problem.

2.2.2. Kinematic Formulation

The kinematic formulation of the problem (2.32) - (2.33) is derived on the ground of the theory of duality. The Lagrange functional in this case has the form:

$$
F(C,\sigma,R,\lambda,\nu,\dot{u}) = \int_V \Lambda C \, dv - \int_V \lambda[\, C - f(\sigma)\,]\,dv \quad -
$$

(2.34)

$$
- \int_V \nu C \, dv - \int_V \dot{u}^T \cdot \nabla\sigma dv - \int_{S_u} \dot{u}^T \cdot (N\sigma - R)\,ds \quad -
$$

$$
- \int_{S_p} \dot{u}^T \cdot (N\sigma - P)\,ds.
$$

The variations of the functional with respect to the variables of the primal problem (C, σ, R) when set equal to zero, lead to the conditions of the dual problem:

$$
\left.
\begin{array}{r}
\Lambda - \lambda - \nu = 0 \\
\lambda \geqslant 0, \quad \nu \geqslant 0 \\
\lambda \dfrac{\partial f(\sigma)}{\partial \sigma} - \nabla^T \dot{u} = 0
\end{array}
\right\} \text{ in } V,
$$

(2.35)

$$
\left.
\begin{array}{r}
\dot{u} = 0 \quad \text{on } S_u.
\end{array}
\right\}
$$

where the variation with respect to σ is based on (2.11). By substituting (2.35) in (2.34) we obtain the kinematic formulation of the body design problem:

$$\max \left\{ \int\limits_{S_p} \dot{u}^T \cdot P ds + \left[\int\limits_V \lambda f(\sigma) dv - \int\limits_V \left[\lambda \frac{\partial f(\sigma)}{\partial \sigma} \right]^T \cdot \sigma dv \right] \right\} \qquad (2.36)$$

subject to

$$\left. \begin{array}{r} \lambda \dfrac{\partial f(\sigma)}{\partial \sigma} - \nabla^T \dot{u} = 0 \\[2mm] \lambda \geqslant 0 \\[1mm] \lambda \leqslant \Lambda \end{array} \right\} \text{ in } V, \\ \left. u = 0 \quad \text{ on } S_u . \right\} \qquad (2.37)$$

The unknown values in this problem are the vector fields σ and $\dot{u}$ as well as the scalar field λ . Now let us turn to the physical sense of the problem.

The first term of the objective function defines the external power, the term in the square brackets at the optimal solution is zero. Thus, the expression (2.36) determines the maximum value of the external power. The first and the fourth equations in (2.37) express the kinematically possible field of the displacement velocities, the remaining inequalities stand for the admissible one (the scalar field of the multipliers is determined within the upper and lower bounds).

Thus, the conditions (2.37) taken together define the kinematically admissible field of the displacement velocities and the problem (2.36) - (2.37) obeys the following extremum problem:

Of all kinematically admissible fields of displacement velocities that one is actual at simple plastic failure, which corresponds to the maximum value of external power.

This principle will be referred to as the kinematic theorem of simple plastic failure.

2.2.3. Theorems of Duality

Since the mathematical models (2.32) - (2.33) and (2.36) - (2.37) are dual, the static and kinematic theorems of plastic failure are dual as well. Now the theorems of duality for these will be formulated.

According to the first theorems of the theory of duality the objective functions for the optimal solutions of dual problems are equal, i.e.

(2.38)
$$\int_V \Lambda C^* \, dv = \int_{S_p} \dot{u}^{*T} \cdot P \, ds.$$

This equation shows that the plastic dissipation is equal to the external power for the actual distribution of the plasticity constant and of the displacement velocities. This will be referred to as the first theorem of duality of the body design problem for a rigid-plastic body.

The second theorem (additional orthogonality condition) has the form:

(2.39)
$$\lambda^*[C - f(\sigma^*)] = 0.$$

This condition is similar to that of the associated plastic flow rule (2.16).

The generalized Lagrange problem for the dual pair obtained may be written as:

(2.40)
$$\left.\begin{array}{r}
C - f(\sigma) \geqslant 0 \\
C \geqslant 0 \\
\nabla \sigma = 0 \\
\lambda \dfrac{\partial f(\sigma)}{\partial \sigma} - \nabla^T \dot{u} = 0 \\
\lambda \geqslant 0 \\
\lambda \leqslant \Lambda \\
\lambda > 0, \text{ subject to } C - f(\sigma) = 0 \\
\lambda = 0, \text{ subject to } C - f(\sigma) > 0
\end{array}\right\} \text{ in } V,$$

$$\left.\begin{array}{r}
N\sigma - R = 0 \\
\dot{u} = 0
\end{array}\right\} \text{ on } S_u,$$

$$N\sigma = P \quad \text{ on } S_p.$$

The field C, σ, $\dot{u}$ and λ satisfying the conditions (2.40) are the solution of the optimization problem for a rigid-plastic body. It is obvious, that the calculation of a separate dual pair is less complicated than the solution of the problem (2.40).

2.2.4. Proof of Static Theorem of Simple Plastic Failure

Similarly to the load design problem the Kuhn-Tucker theorem is used to prove the primal theorem leading to the mathematical models of the body design problem.

The stress field σ^* satisfying the conditions of the statically admissible field (2.33) and corresponding to the minimum value of the plastic dissipation is denoted by an asterisk. Let the actual field σ satisfy only the conditions (2.33). The distribution of the plasticity constant C for the body is assumed to be the same. Then, the Kuhn-Tucker theorem for the Lagrange functional (2.34) has the form:

$$F(C,\sigma,R,\lambda^*,\nu^*,\dot{u}^*) - F(C^*,\sigma^*,R^*,\lambda^*,\nu^*,\dot{u}^*) \geqslant 0. \qquad (2.41)$$

By substituting (2.34) in (2.41), we obtain:

$$\int_V \lambda^* [\, f(\sigma) - f(\sigma^*)\,] \geqslant 0. \qquad (2.42)$$

Since the expression (2.42) must be positive, we have:

$$c - f(\sigma) \leqslant 0.$$

which does not correspond to the primal condition of plasticity. Hence, as in the case of the load design problem, the relations (2.42) may be satisfied only as equality. It follows that

$$\sigma = \sigma^*. \qquad (2.43)$$

i.e. the stress field is actual at plastic failure if it minimizes the value of plastic dissipation. The theorem has been proved.

2.2.5. One-Parametrical Problem

The mathematical models for the one-parametrical body design problem will be derived.

Let the distribution of the plasticity constant of the material in the volume V be prescribed by:

$$C = C^\circ \gamma. \qquad (2.44)$$

where C° is the required parameter, and γ is the prescribed scalar field of the distribution of the plasticity constant. The plastic dissipation will be:

$$(2.45) \qquad\qquad\qquad \dot{D} = C^\circ \int_V \lambda\gamma dv.$$

By fixing the multiplier $\int_V \lambda\gamma dv \equiv 1$, we obtain the dual pair of problems enabling the parameter of the plasticity constant to be determined:

a) **Static Formulation**

$$(2.46) \qquad\qquad\qquad\qquad\qquad \min C^\circ$$

subject to

$$(2.47) \qquad \left.\begin{array}{rcll} C^\circ\gamma - f(\sigma) & \geqslant & 0 & \\ C^\circ & \geqslant & 0 & \text{in } V, \\ \nabla\sigma & = & 0 & \\ N\sigma - R & = & 0 & \text{on } S_u, \\ N\sigma & = & P & \text{on } S_p. \end{array}\right\}$$

b) **Kinematic Formulation**

$$(2.48) \qquad \max\left\{\left[\int_{S_p} \dot{u}^T \cdot P ds + \left[\int_V \lambda f(\sigma) dv - \int_V \left[\lambda \frac{\partial f(\sigma)}{\partial\sigma}\right]^T \cdot \sigma dv\right]\right\}$$

subject to

$$(2.49) \qquad \left.\begin{array}{rcll} \lambda \dfrac{\partial f(\sigma)}{\partial\sigma} - \nabla^T \dot{u} & = & 0 & \\ \lambda & \geqslant & 0 & \text{in } V, \\ \int_V \lambda\gamma dv & \leqslant & 1 & \\ \dot{u} & = & 0 & \text{on } S_u. \end{array}\right\}$$

This dual pair of problems obeys the theorems of duality. According to the first one, we have:

$$C^{\circ *} = \int_{S_p} \dot{u}^{*T} \cdot P ds.\tag{2.50}$$

This expression shows that the parameter of the plasticity constant corresponding to the actual field of the displacement velocities is equal to the external power. If we multiply the conditions $\int_V \lambda \gamma dv \leqslant 1$ of (2.49) C° this equation will express the fixed plastic dissipation.

The generalized Lagrange problem for the dual pair of one-parametrical problems will be:

$$
\left.
\begin{array}{r}
C^{\circ}\gamma - f(\sigma) \geqslant 0 \\[4pt]
C^{\circ} \geqslant 0 \\[4pt]
\nabla \sigma = 0 \\[4pt]
\lambda \dfrac{\partial f(\sigma)}{\partial \sigma} - \nabla^{T}\dot{u} = 0 \\[4pt]
\lambda \geqslant 0 \\[4pt]
\int_V \lambda \gamma dv \leqslant 1 \\[4pt]
\lambda > 0, \text{ subject to } C^{\circ}\rho - f(\sigma) = 0 \\[4pt]
\lambda = 0, \text{ subject to } C^{\circ}\rho - f(\sigma) > 0 \\[8pt]
\left.
\begin{array}{r}
N\sigma - R = 0 \\
\dot{u} = 0
\end{array}
\right\} \text{ on } S_u, \\[8pt]
N\sigma = P \quad \text{ on } S_p.
\end{array}
\right\}
\left.
\begin{array}{l}
\\ \\ \text{in } V, \\ \\ \\ \\ \\ (2.51) \\ \\ \\ \\ \\ \\
\end{array}
\right.
$$

2.2.6. Comparison of Conditions of Optimization and One-Parametrical Problems

Comparison between the conditions of the one-parametrical body design problem (2.51) and those of the optimization problem (2.40) clearly shown that the difference lies in the constraints on the scalar field of the multipliers λ. In the optimization problem and in the one parametrical one the constraints are $\lambda \leqslant \lambda$

and $\oint_V \lambda\gamma dv \leqslant 1$. respectively. It can easily be shown that $C^\circ > 0$ satisfies the latter as equality on the ground of the additional orthogonality conditions which implies that:

$$C^{\circ*}\left[\int_V \lambda^* \gamma dv - 1\right] = 0.$$

Since $C^{\circ*} > 0$(otherwise the body does not exist), we have:

$$\int_V \lambda^* \gamma dv = 1$$

Thus, the conditions $\lambda \leqslant \Lambda$ and $\int_V \lambda\gamma dv = 1$ are a form of normalization and permit the field of the displacement velocities to be determined to within a constant multiplier. The scalar field of the weight multipliers of the optimality criterion Λ defines the upper bound for every component of the strain rates in the volume V and are of kinematic origin. In the one-parametrical problem the constraint for the strain rate has an integral form. It is obvious, that the first constraint is tighter with respect to the variables, but at the same time it makes possible a free selection of the constraints and thus permits an optimal distribution of the plasticity constant to be obtained.

2.3. Geometrical Interpretation of Load Design and Body Design Problems

One-parametrical problems for a rigid-plastic body will be considered from a geometrical point of view. Let the stress field of the body be determined by the two-dimensional vector $\sigma \equiv (\sigma_1 \sigma_2)^T$ then one-parametrical problems may be plotted in a three-dimensional space (Fig. 2.1 and Fig. 2.2).

In the load design problem, the admissible stress field is enclosed by a cylinder the base of which is the curve $f(\sigma) = C$. Statically possible field expressed by the equation $N\sigma - P^\circ\eta = 0$ is the plane passing through the origin of coordinates. (Fig. 2.1a). Every point of this plane inside the cylinder or on its surface will correspond to the conditions of the statically admissible field. The point max P° is the highest ordinate on the intersection curve of the cylinder and the plane. The relevant coordinate σ^* on the stress plane expresses the actual stress field at plastic failure. If the yield function is essentially convex, as shown in Fig. 2.1a, then max P° will correspond to the single coordinate σ^* and to the single value of the gradient on the yield surface, i.e. to the single value of the strain rate. If the yield function is

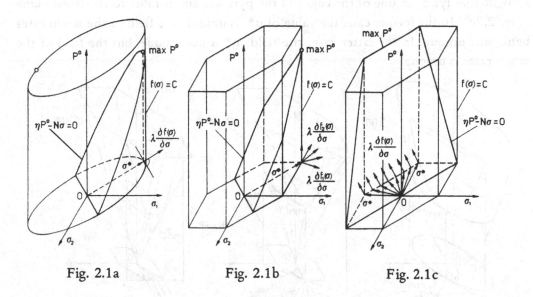

Fig. 2.1a   Fig. 2.1b   Fig. 2.1c

piecewise linear, the conditions for the admissible field are restricted by a prism whose generatrices are parallel to the axis of ordinates. In this case max P^o is obtained on the edge or on the straight line lying on one of the one-side surfaces and parallel to the stress plane. In the former case, there exists a unique value of σ^* the direction of the gradient being non-unique, i.e. the value of the strain rate at the given point is not unique too (Fig. 2.1b). In the latter case, we have a set of vectors σ^* corresponding to max P^o; nevertheless, the entire set corresponds to a unique value of the strain rate (Fig. 2.1c).

In the one-parametrical body design problem the conditions for the permissible field $C^o\rho - f(\sigma) = 0$ are restricted by a paraboloid, whose cross-sections are parallel to the stress plane and express the yield condition for the corresponding values of C^o (Fig. 2.2a). The equilibrium equations are expressed by the plane parallel to the axis of ordinates, but not passing through the origin of coordinates. The intersection line of this plane and the paraboloid is a curve whose lowest point expresses min C^o. The coordinate corresponding to this point on the stress plane expresses the actual field σ^*. If the yield function $f(\sigma)$ is essentially convex, this point is unique, and the curve expressing the yield conditions will have in it a unique gradient, i.e. a unique value of the strain rate. The yield function being piecewise linear, min C° may be obtained on the edge of the pyramid (Fig. 2.2b) or on the straight line lying on one of the edges of the pyramid and parallel to the stress plane (Fig. 2.2c). In the former case, the value of σ^* is unique, the field of the strain rates being not unique. In the latter case, the field σ^* is not unique, but the field of the strain rates is unique.

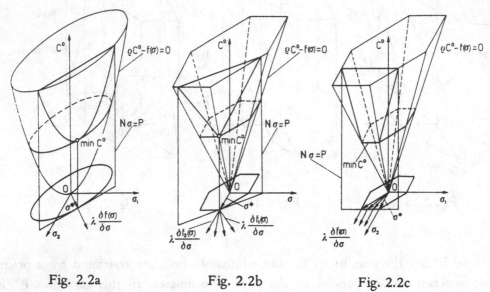

Fig. 2.2a   Fig. 2.2b   Fig. 2.2c

Fig. 2.1 and Fig. 2.2 clearly show that the one-parametrical load design and body design problems are trivial. Both max P° and min C° correspond to the points on the intersection curve of the plane of the equilibrium equations and the surface of the admissible field. In the former case, the plane passes through the origin of coordinates, and the surface of the admissible field is enclosed by the cylinder whose generatrices are parallel to the axis of ordinates. In the latter case, the plane of the equilibrium equations is parallel to the axis of ordinates, and the surface of the permissible field passes through the origin of coordinates. In both cases, there is a yield curve on the stress plane, where the ordinate σ^* expresses the actual stress field corresponding to max P° or min C°.

The geometrical interpretation of one-parametrical problems may be applied to optimization problems as well. In this case the load design and body design problems will not be trivial, since the hyperplanes for the optimality criteria of these types of problems are different. Nevertheless, the conclusions concerning the actual fields of stresses and strain rates will be the same as in one-parametrical problems.

Table 2.1 shows the relationship between various formulations of the problems for rigid-plastic bodies and the objective of every type of problem.

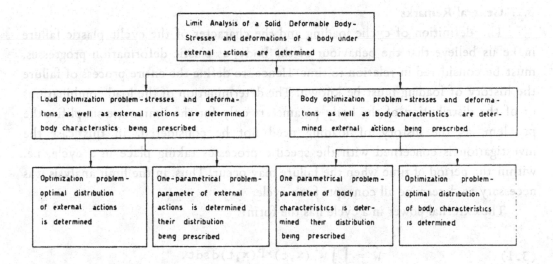

III. ELASTIC-PLASTIC BODY UNDER CYCLIC LOADING

3.1. General Remarks

The definition of cyclic loading and the character of the cyclic plastic failure make us believe that the behaviour of the body, as plastic deformation progresses, must be considered in relation to time. Hence, to define the entire process of failure the history of loading must be known. The determination of the load combinations or of the distribution of the body parameters only for the limit state simplifies the problem, as the history of loading need not be considered. In this case the investigation is concerned with the specific processes taking place in a cycle, i.e. within the period of time when the failure may occur. Thus, in the limit analysis it is necessary to determine all concepts for a cycle.

The external power in a cycle has the form:

$$(3.1) \qquad \dot{W} = \int_\tau \int_{S_p} \dot{u}^T(x,t) \cdot P(x,t)\, ds\, dt .$$

where the displacement velocity and the loading are the functions of the coordinates and time determined by the extremum values independent of time. Since in the load design problem the displacement velocities are fixed, we can use a well known property of the definite integral (mean-value theory) permitting the subintegral function to be expressed by corresponding extremum values:

$$(3.2.) \qquad \dot{W} = \int_{S_p} \dot{u}^{+T} \cdot P^+\, ds + \int_{S_p} u^{-T} \cdot P^-\, ds .$$

Similarly, the plastic dissipation may be expressed by:

$$\dot{D} = \int_\tau \int_V \dot{\epsilon}^T(x,t) \cdot \sigma(x,t)\, dt\, dv = \int_V \dot{\epsilon}^{+T} \cdot \sigma^+\, dv \ + $$

$$(3.3)$$

$$+ \int_V \dot{\epsilon}^{-T} \cdot \sigma^-\, dv = \int_V (\lambda^+ + \lambda^-)\, C\, dv .$$

The duration of a cycle in (3.2) and (3.3) is assumed to be $\tau \equiv 1$.

In case of monotonically increasing loading, the stress field is determined by the field of the actual stresses σ, while in the case of cyclic loading it may be of two kinds, i.e. it is determined either by the maximum or the minimum values of the stresses. These extremum values of the actual stresses may be expressed as the sum of extremum "elastic" and residual stresses independent of time. Since the extremum values of elastic stresses, determined on the ground of the boundary conditions of elasticity theory, are assumed to be prescribed completely or in the form of the linear dependence on the scalar weight multiplier, then the residual stresses are the main values to be found. With this aim in view the basic extremum theorems are formulated, and the admissible and statically possible stress fields are determined. The field of residual stresses is admissible if, when added to the extremum "elastic" stresses, it satisfies the yield conditions, i.e.:

$$\left. \begin{aligned} f(\sigma^+ + \rho) &\leqslant c, \\ f(-\sigma^- - \rho) &\leqslant c. \end{aligned} \right\} \tag{3.4}$$

The field of residual stresses is statically possible if it satisfies the differential equilibrium equations and the primal boundary conditions. The term "residual" itself implies that the stresses are self-balanced, i.e. $N\rho$ is equal to zero on the surface.

The field of residual stresses is statically admissible, if it satisfies the conditions both for the admissible and statically possible fields.

3.2. Load Design Problem

The purpose of the problem is to determine the limit loads corresponding to the optimality criterion as well as the stress and strain fields at cyclic plastic failure.

3.2.1. Static Formulation

The mathematical model of the load design problem is derived on the ground of the extremum principles referred to as the static theorem of limit cyclic load.

Of all statically admissible fields of residual stresses at cyclic plastic failure, that one is actual which corresponds to the maximum value of external power in a cycle.

The mathematical model corresponding to this theorem is:

(3.5)
$$\max \left\{ \int_{S_p} \dot{u}^{+T} \cdot P^+ \, eds + \int_{S_p} \dot{u}^{-T} \cdot P^- \, eds \right\}$$

subject to:

(3.6)
$$\left. \begin{array}{c} f(\sigma^+ + \rho) \leqslant C \\ f(-\sigma^- - \rho) \leqslant C \end{array} \right\} \text{ in } V,$$
$$\left. \begin{array}{r} \nabla \rho = 0 \\ N\rho - r = 0 \quad \text{on } S_u, \\ N\rho = 0 \\ P^+ \geqslant 0, \; P^- \geqslant 0 \end{array} \right\} \text{ on } S_p.$$

where σ^+ and σ^- are determined through (1.12).

Since the displacement velocities are not included in the conditions of the problem, the minimization of (3.5) is possible at any fixed values of $\dot{u}^+$ and $\dot{u}^-$ including $\dot{u}^+ \equiv T^+$ and $\dot{u}^- \equiv T^-$ where T^+ and T^- are the vector fields of the weight multipliers of the optimality criterion for cyclic loading. Thus, we obtain the mathematical model of the load design problem in the static formulation:

(3.7)
$$\max \left\{ \int_{S_p} T^{+T} \cdot P^+ eds + \int_{S_p} T^{-T} \cdot P^- eds \right\}$$

subject to

(3.8)
$$\left. \begin{array}{c} f(\sigma^+ + \rho) \leqslant C \\ f(-\sigma^- - \rho) \leqslant C \end{array} \right\} \text{ in } V,$$
$$\left. \begin{array}{r} \nabla \rho = 0 \\ N\rho - r = 0 \quad \text{on } S_u, \\ N\rho = 0 \\ P^+ \geqslant 0, P^- \geqslant 0 \end{array} \right\} \text{ on } S_p.$$

This is the functional analogue of a convex programming problem, i.e. the linear functional (3.7) is maximized on the convex set (3.8). By solving the problem we obtain the vector field of the residual stresses ρ and the scalar fields of the upper and lower bounds of the load P^+ and P^-. These enable the vector fields of the extremum "elastic" stresses to be obtained through (1.12), which, when added to the residual stresses, result in the actual stresses at cyclic plastic failure.

Note that there exist two values of every component of the actual stresses at the point under consideration, corresponding to the most dangerous combinations of loading in a cycle. If at this point both yield conditions are satisfied as equalities, then we have to do with "variable plasticity". If only one condition is satisfied as equality, then "progressive" failure occurs. Nevertheless, these conclusions are based on the solution of the problem and can by no means be taken as the grounds for formulating the very problem.

3.2.2. Kinematic Formulation

To derive the kinematic formulation of the problem, the Lagrange functional is written as follows:

$$
\begin{aligned}
F(P^+ \; P^-, \rho, r, \lambda^+, \lambda^-, \dot{w}, \nu^+, \nu^-) = & \int_{S_p} T^{+T} \cdot P^+ e ds \quad + \\
& + \int_{S_p} T^{-T} \cdot P^- e ds + \int_V \lambda^+ [C - f(\sigma^+ + \rho)] dv + \int_V \lambda^- [C - \\
& - f(- \sigma^- - \rho)] dv + \int_V \dot{w}^T \cdot \nabla \rho \, dv + \int_{S_u} \dot{w}^T \cdot (N\rho - r) ds + \\
& + \int_{S_p} \dot{w}^T \cdot N\rho \, ds + \int_{S_p} \nu^{+T} \cdot P \; e ds + \int_{S_p} \nu^{-T} \cdot P^- e ds.
\end{aligned}
\tag{3.9}
$$

The variations of the functional with respect to $P^+_;P^-$, ρ and r are set equal to zero, leading to the conditions of o dual problem; then varying the functional with respect to σ^+ and σ^- it should be mentioned that these values depend on P^+ and P^-. Thus, we have:

$$-\lambda^+ \frac{\partial f(\sigma^+ + \rho)}{\partial \rho} - \lambda^- \frac{\partial f(-\sigma^- - \sigma)}{\partial \rho} + \nabla^T \dot{w} = 0$$

$$\lambda^+ \geqslant 0, \quad \lambda^- \geqslant 0 \quad \Biggr\} \text{ in } V,$$

$$T^+ - \int_V \lambda^+ \frac{\partial f(\sigma^+ + \rho)}{\partial P^+} dv - \int_V \lambda^- \frac{\partial f(-\sigma^- - \rho)}{\partial P^+} dv + \nu^+ = 0$$

$$T^- - \int_V \lambda^+ \frac{\partial f(\sigma^+ + \rho)}{\partial P^-} dv - \int_V \lambda^- \frac{\partial f(-\sigma^- - \rho)}{\partial P^-} dv + \nu^- = 0 \quad \Biggr\} \text{ on } S_p,$$

$$\nu^+ \geqslant 0, \quad \nu^- \geqslant 0$$

(3.10) $$\dot{w} = 0 \text{ Ha } S_u.$$

Substituting (3.10) in (3.9) and keeping in mind that the residual stresses at the end of a cycle become the same as the initial ones at the beginning of it, i.e. the plastic dissipation at the residual velocities $\int_V \rho^T \cdot \dot{\xi} dv = 0$ and taking into account the relations:

(3.11)
$$\frac{\partial f(\sigma^+ + \rho)}{\partial P^+} = \omega^{+T} \frac{\partial f(\sigma^+ + \rho)}{\partial \rho},$$

$$\frac{\partial f(-\sigma^- - \rho)}{\partial P^+} = \omega^{-T} \frac{\partial f(-\sigma - \rho)}{\partial \rho},$$

$$\frac{\partial f(\sigma^+ + \rho)}{\partial P^-} = -\omega^{-T} \frac{\partial f(\sigma^+ + \rho)}{\partial \rho},$$

$$\frac{\partial f(-\sigma^- - \rho)}{\partial P^-} = -\omega^{+T} \frac{\partial f(-\sigma^- - \rho)}{\partial \rho}$$

and

$$\left[\lambda^+ \frac{\partial f(\sigma^+ + \rho)}{\partial \rho}\right]^T \cdot \int_{S_p} (\omega^+ P^+ e - \omega^- P^- e) ds = \left[\lambda^+ \frac{\partial f(\sigma^+ + \rho)}{\partial \rho}\right]^T \cdot \sigma^+ = \dot{\epsilon}^{+T} \cdot \sigma^+,$$

$$\left[\lambda^- \frac{\partial f(-\sigma^- - \rho)}{\partial \rho}\right]^T \cdot \int_{S} (\omega^- P^+ e - \omega^+ P^- e) ds = \left[\lambda^- \frac{\partial f(-\sigma^- - \rho)}{\partial \rho}\right]^T \cdot \sigma^- = \dot{\epsilon}^{-T} \cdot \sigma^-;$$

(3.12)

we obtain the mathematical model of the load design problem in the kinematic formulations:

$$\min \left\{ \int_V \left[\lambda^+ \frac{\partial f(\sigma^+ + \rho)}{\partial \rho} \right]^T \cdot \sigma^+ \, dv + \int_V \left[\lambda^- \frac{\partial f(-\sigma^- - \rho)}{\partial \rho} \right]^T \cdot \sigma^- dv \right. +$$

$$+ \left. \int_V \lambda^+ [c - f(\sigma^+ + \rho)] \, dv + \int_V \lambda^- [c - f(-\sigma^- - \rho)] \, dv \right\} \qquad (3.13)$$

subject to

$$\left. \begin{array}{l} - \lambda^+ \dfrac{\partial f(\sigma^+ + \rho)}{\partial \rho} - \lambda^- \dfrac{\partial f(-\sigma^- - \rho)}{\partial \rho} + \nabla^T \dot{w} = 0 \\ \qquad\qquad \lambda^+ \geqslant 0, \quad \lambda^- \geqslant 0 \end{array} \right\} \text{ in } V,$$

$$\left. \begin{array}{l} \displaystyle\int_V \lambda^+ \dfrac{\partial f(\sigma^+ + \rho)}{\partial P^+} \, dv + \int_V \lambda^- \dfrac{\partial f(-\sigma^- - \rho)}{\partial P^+} \, dv \geqslant T^+ \\[4mm] \displaystyle\int_V \lambda^+ \dfrac{\partial f(\sigma^+ + \rho)}{\partial P^-} \, dv + \int_V \lambda^- \dfrac{\partial f(-\sigma^- - \rho)}{\partial P^-} \, dv \geqslant T^- \end{array} \right\} \text{ on } S_p, \quad (3.14)$$

$$\dot{w} = 0 \text{ on } S_u$$

Now let us explain the physical sense of the problem (3.13) (3.14). If we denote:

$$\dot{\epsilon}^+ = \lambda^+ \frac{\partial f(\sigma^+ + \rho)}{\partial \rho}, \qquad \dot{\epsilon}^- = \lambda^- \frac{\partial f(-\sigma^- - \rho)}{\partial \rho},$$

$$\dot{\xi} = \dot{\epsilon}^+ + \dot{\epsilon}^- . \qquad\qquad (3.15)$$

then the first and the last conditions in (3.14) correspond to the kinematically possible field of the velocities of the residual displacements $\dot{w}$. The remaining conditions in (3.14) are the constraints in the form of inequalities and define the kinematically admissible field of the velocities of the residual displacements in a certain cycle.

Now let us explain the physical sense of the objective function (3.13).

The first two integrals express the plastic dissipation in a cycle, the third and the fourth terms of (3.13) are zero according to the additional orthogonality condition.

Thus, the objective function expresses the plastic dissipation in a cycle, and the problem (3.11) (3.10) corresponds to the following theorem:

Of all kinematically admissible fields of velocities of residual displacements at cyclic plastic failure, that one is actual which corresponds to the minimum value of plastic dissipation.

This is the kinematic theorem of cyclic limit load.

3.2.3. Theorems of Duality

Since the mathematical model (3.13 - (3.14) is dual to the problem (3.7) - (3.8), the static and kinematic theorems should be regarded as dual.

Corresponding to the first theorem of duality, we have:

$$\int_{S_p} T^{+T} \cdot P^{+*} e \, ds + \int_{S_p} T^{-T} \cdot P^{-*} e \, ds = \int_V \left[\lambda^{+*} \frac{\partial f(\sigma^{+*} + \rho^*)}{\partial \rho} \right]^T \cdot \sigma^{+*} dv +$$
$$+ \int_V \left[\lambda^{-*} \frac{\partial f(-\sigma^{-*} - \rho^*)}{\partial \rho} \right]^T \cdot \sigma^{-*} dv.$$

(3.16)

From a physical point of view this means, that the external power and the plastic dissipation in a cycle are equal. The relation (3.16) will be referred to as the first theorem of duality of the load-design problem of cyclic loading. The second theorem of duality has the form:

(3.17)
$$\begin{aligned} \lambda^{+*} [C - f(\sigma^{+*} + \rho^*)] &= 0, \\ \lambda^{-*} [C - f(-\sigma^{-*} - \rho^*)] &= 0. \end{aligned}$$

These conditions are similar to those of the associated plastic flow rule at cyclic failure:

$$
\left.
\begin{aligned}
&\lambda^+ > 0, \text{ subject to } f(\sigma^+ + \rho) = C, \\
&\lambda^+ = 0, \text{ subject to } f(\sigma^+ + \rho) < C, \\
&\lambda^- > 0, \text{ subject to } f(-\sigma^- - \rho) = C, \\
&\lambda^- = 0, \text{ subject to } f(-\sigma^- - \rho) < C.
\end{aligned}
\right\} \qquad (3.18)
$$

If for a given point $\lambda^+ > 0$ and $\lambda^- > 0$ then this is the case of "variable plasticity", and if only one of these conditions is satisfied as inequality, then "progressive" failure occurs.

The generalized Lagrange problem will be written for the dual pairs of the problems (3.7) - (3.8) and (3.13) - (3.14). It consists of all conditions of both problems and (3.18):

$$
\left.
\begin{aligned}
&f(\sigma^+ + \rho) \leq C \\
&f(-\sigma^- - \rho) \leq C \\
&\nabla \rho = 0 \\
&-\lambda^+ \frac{\partial f(\sigma^+ + \rho)}{\partial \rho} - \lambda^- \frac{\partial f(-\sigma^- - \rho)}{\partial \rho} + \nabla^T \dot{w} = 0 \\
&\lambda^+ \geq 0, \quad \lambda^- \geq 0. \\
&\lambda^+ > 0, \text{ subject to } f(\sigma^+ + \rho) = C \\
&\lambda^+ = 0, \text{ subject to } f(\sigma^+ + \rho) < C \\
&\lambda^- > 0, \text{ subject to } f(-\sigma^- - \rho) = C \\
&\lambda^- = 0, \text{ subject to } f(-\sigma^- - \rho) < C \\
&\sigma^+ = \int_{-V} (\omega^+ P^+ e - \omega^- P^- e) ds \\
&\sigma^- = \int_{V} (\omega^- P^+ e - \omega^+ P^- e) ds
\end{aligned}
\right\} \text{ in } V, \quad (3.19)
$$

$$P^+ \geqslant 0, \quad P^- \geqslant 0$$

$$(3.19) \quad \left. \begin{array}{l} \displaystyle\int_V \lambda^+ \frac{\partial f(\sigma^+ + \rho)}{\partial P^+}\, dv + \int_V \lambda^- \frac{\partial f(-\sigma^- - \rho)}{\partial P^+}\, dv \geqslant T^+ \\[4mm] \displaystyle\int_V \lambda^+ \frac{\partial f(\sigma^+ + \rho)}{\partial P^-}\, dv + \int_V \lambda^- \frac{\partial f(-\sigma^- - \rho)}{\partial P^-}\, dv \geqslant T^- \\[4mm] N\rho = 0 \\[2mm] \dot{w} = 0, \quad N\rho - r = 0 \quad \text{on } S_u. \end{array} \right\} \text{on } S_p ,$$

The fields $P^+, P^-, \rho, \dot{w}$ and λ^+, λ^- satisfying these conditions are the solution of the load design problem for an elastic-plastic body under cyclic loading.

3.2.4. Proof of Static Theorem of Cyclic Limit Load

The mathematical models (3.7) - (3.8) and (3.13) - (3.14) were derived on the ground of the static theorem of cyclic limit load. Now the theorem is to be proved.

Let there exist some field of residual stresses ρ satisfying the conditions (3.8) but not corresponding to the maximum value of (3.7). The solution ρ^* of the extremum problem (3.7) - (3.8) is denoted by an asterisk. The fields ρ^* and ρ are assumed to correspond to only one field of cyclic limit load. According to the Kuhn-Tucker theorem, the Lagrange functional (3.9) must satisfy the inequality:

$$(3.20) \quad \begin{array}{l} F(P^{+*}, P^{-*}, \rho^*, r^*, \lambda^{+*}, \lambda^{-*}, \dot{w}^*, \nu^{+*}, \nu^{-*}) - \\[2mm] - F(P^+, P^-, \rho, r, \lambda^{+*}, \lambda^{-*}, \dot{w}^*, \nu^{+*}, \nu^{-*}) \geqslant 0. \end{array}$$

Substituting the expression (3.9) in (3.10) and taking into consideration that $P^+ = P^{+*}$ and $P^- = P^{-*}$, we obtain:

$$\int_V \lambda^{+*} [f(\sigma^{+*} + \rho) - f(\sigma^{+*} + \rho) \, dv] +$$
$$+ \int_V \lambda^{-*} [f(-\sigma^- - \rho) - f(-\sigma^{-*} - \rho^*)] dv \geqslant 0. \tag{3.21}$$

Since for the optimal solution of the problem (3.7) - (3.8) we have $\lambda^{+*} f(\sigma^{+*}+\rho^*) = \lambda^{+*} C \lambda^{-*} f(-\sigma^{-*}-\rho^*) = \lambda^{-*} C$, expressions in the square brackets of the inequality (3.21) cannot be positive, otherwise the yield conditions are not satisfied. Therefore, these may equal only zero. It follows that:

$$f(\sigma^+ + \rho) = f(\sigma^{+*} + \rho^*),$$
$$f(-\sigma^- - \rho) = f(-\sigma^{-*} - \rho^*)$$

Taking into account that $P^+ = P^{+*}$, $P^- = P^{-*}\sigma^+ = \sigma^{+*}$ $\sigma^- = \sigma^{-*}$ we have :

$$\rho = \rho^*.$$

Hence, of all statically admissible fields of residual stresses the actual one corresponds to the maximum value of the external power in a cycle. The theorem has been proved.

3.2.5. One-Parametrical Problem

The one-parametrical load design problem for cyclic loading will be considered. Let the load on the surface S_p be prescribed by:

$$P^+ = P^\circ \eta^+ , \quad P^- = P^\circ \eta^-, \tag{3.22}$$

where P° is the load parameter and η^+ and η^- are the vector fields determining the distribution of the upper and lower bounds of cyclic loading. Thus, by varying the parameter P°, the bounds for all external forces may be narrowed or extended simultaneously. Now the external power in a cycle has the form:

$$W = \int_{S_p} P^\circ \dot{u}^{+T} \cdot \eta^+ ds + \int_{S_p} P^\circ \dot{u}^{-T} \cdot \eta^- ds =$$
$$= P^\circ \int_{S_p} (\dot{u}^{+T} \cdot \eta^+ + \dot{u}^{-T} \cdot \eta^-) ds. \tag{3.23}$$

If the behaviour of the body is assumed to be elastic then the extremum values of the stresses will be:

$$
\begin{aligned}
\sigma^+ &= \int_{S_p} (\omega^+ P^\circ \eta^+ - \omega^- P^\circ \eta^-)\,ds = P^\circ \int_{S_p} (\omega^+ \eta^- - \\
&\qquad\qquad\qquad\qquad - \omega^- \eta^-)\,ds = P^\circ \Omega^+, \\
\sigma^- &= \int_{S_p} (\omega^- P^\circ \eta^+ - \omega^+ P^\circ \eta^-)\,ds = P^\circ \int_{S_p} (\omega^- \eta^+ - \\
&\qquad\qquad\qquad\qquad - \omega^+ \eta^-)\,ds = P^\circ \Omega^-
\end{aligned}
$$

(3.24)

Fixing the multiplier $\int_{S_p}(\dot{u}^{+T}\cdot\eta^+ + \dot{u}^{-T}\cdot\eta^-)\,ds \equiv 1$ we obtain the mathematical model of the one-parametrical problem:

a) **Static Formulation**

(3.25)
$$\max P^\circ$$

subject to

$$
\left.
\begin{aligned}
f(\sigma^+ + \rho) &\leq c \\
f(-\sigma^- - \rho) &\leq c \\
\nabla\rho &= 0
\end{aligned}
\right\} \text{ in } V,
\qquad
\left.
\begin{aligned}
& \\
& \\
&
\end{aligned}
\right\}
$$

(3.26)
$$
\begin{aligned}
N\rho - r &= 0 && \text{on } S_u, \\
N\rho &= 0, \quad P^\circ \geq 0 && \text{on } S_p;
\end{aligned}
$$

b) **Kinematic Formulation**

$$
\min \left\{ \int_V \left[\lambda + \frac{\partial f(\sigma^+ + \rho)}{\partial\rho}\right]^T \cdot \sigma^+\,dv + \int_V \left[\lambda - \frac{\partial f(-\sigma^- - \rho)}{\partial\rho}\right]^T \cdot \sigma^-\,dv + \right.
$$

$$+ \int_V \lambda^+ [C - f(\sigma^+ + \rho)] dv + \int_V \lambda^- [C - f(-\sigma^- - \rho)] dv \Big\} \qquad (3.27)$$

subject to

$$\left. \begin{array}{c} - \lambda^+ \dfrac{\partial f(\sigma^+ + \rho)}{\partial \rho} - \lambda^- \dfrac{\partial f(-\sigma^- - \rho)}{\partial \rho} + \nabla^T \dot{w} = 0 \\[2mm] \lambda^+ \geqslant 0, \quad \lambda^- \geqslant 0 \end{array} \right\} \text{ in V,} $$

$$\int_V \left[\lambda^+ \frac{\partial f(\sigma^+ + \rho)}{\partial P^\circ} \right] dv + \int_V \left[\lambda^- \frac{\partial f(-\sigma^- - \rho)}{\partial P^\circ} \right] dv \geqslant 1,$$

$$\dot{w} = 0 \quad \text{on } S_u.$$

$$(3.28)$$

The objective function (3.27), as in the problem (3.13) - (3.14), expresses the plastic dissipation because the third and the fourth terms of it for the optimal solution are zero.

The integral constraint in (3.28) may be looked upon as fixed power, as the following relations are valid:

$$\int_V \lambda^+ \frac{\partial f(\sigma^+ + \rho)}{\partial P^\circ} dv + \int_V \lambda^- \frac{\partial f(-\sigma^- - \rho)}{\partial P^\circ} dv = \int_V (\dot{\epsilon}^{+T} \Omega^+ + \dot{\epsilon}^{-T} \Omega^-) dv =$$

$$= \int_V \int_{S_p} [\dot{\epsilon}^{+T} \cdot (\omega^+ \eta^+ - \omega^- \eta^-) + \dot{\epsilon}^{-T} \cdot (\omega^- \eta^+ - \omega^+ \eta^-)] dv ds =$$

$$= \int_V \int_{S_p} [(\dot{\epsilon}^{+T} \omega^+ + \dot{\epsilon}^{-T} \omega^-) \cdot \eta^+ - (\dot{\epsilon}^{+T} \omega^- + \dot{\epsilon}^{-T} \omega^+) \cdot \eta^-] dv ds =$$

$$= \int_{S_p} (u^{+T} \cdot \eta^+ + u^{-T} \cdot \eta^-) ds.$$

Thus, the integral constraint in the kinematic formulation of the problem may be substituted by:

$$\int\limits_{S_p} (\dot{u}^{+\,T}\cdot\eta^+ + \dot{u}^{-\,T}\cdot\eta^-)ds \geqslant 1.$$

Nevertheless, the substitution does not seem to be justified as the explicit form of the variables $\dot{u}^+$ and $\dot{u}^-$ is not included in (3.27) - (3.28).

The first theorem of duality yields:

$$(3.29) \quad P^\circ = \int\limits_V \left[\lambda^{+*}\,\frac{\partial f(\sigma^{+*} + \rho^*)}{\partial\rho}\right]^T \cdot \sigma^{+*}\,dv \; +$$

$$+ \int\limits_V \left[\lambda^{-*}\,\frac{\partial f(-\sigma^{-*} - \rho^*)}{\partial\rho}\right]^T \cdot \sigma^{-*}\,dv.$$

The generalized Lagrange problem for the one-parametrical load design problem has the form:

$$(3.30) \quad \left.\begin{array}{l} f(\sigma^+ + \rho) \leqslant C \\[4pt] f(-\sigma^- - \rho) \leqslant C \\[4pt] \qquad \nabla\rho = 0 \\[8pt] -\lambda^+\,\dfrac{\partial f(\sigma^+ + \rho)}{\partial\rho} - \lambda^-\,\dfrac{\partial f(-\sigma^- - \rho)}{\partial\rho} + \nabla^T\dot{w} = 0 \\[8pt] \lambda^+ \geqslant 0, \quad \lambda^- \geqslant 0 \\[6pt] \lambda^+ > 0, \text{ subject to } f(\sigma^+ + \rho) = C \\[4pt] \lambda^+ = 0, \text{ subject to } f(\sigma^+ + \rho) < C \\[4pt] \lambda^- > 0, \text{ subject to } f(-\sigma^- - \rho) = C \\[4pt] \lambda^- = 0, \text{ subject to } f(-\sigma^- - \rho) < C \\[6pt] \sigma^+ = P^\circ\!\!\int\limits_{S_p} (\omega^+\eta^+ - \omega^-\eta^-)ds = P^\circ\Omega^+ \\[8pt] \sigma^- = P^\circ\!\!\int\limits_{S_p} (\omega^-\eta^+ - \omega^+\eta^-)ds = P^\circ\Omega^- \end{array}\right\} \text{ in } V,$$

$$\left. \begin{array}{c} \int\limits_{V} \lambda^+ \dfrac{\partial f(\sigma^+ + \rho)}{\partial P^\circ} \, dv + \int\limits_{V} \lambda^- \dfrac{\partial f(-\sigma^- - \rho)}{\partial P^\circ} \, dy \geq 1 \\ N\rho = 0, \qquad P^\circ \geq 0 \\ \dot{w} = 0, \qquad N\rho - r = 0 \ \ \text{on} \ S_u \end{array} \right\} \text{on} \ S_p, \quad (3.30)$$

The fields $P^\circ, \rho, \lambda^+, \lambda^-, \dot{w}$, satisfying all conditions (3.30) are the solution of the problem determining the parameter cyclic loading for an elastic-plastic body.

3.2.6. Comparison of Conditions of Optimization and One-Parametrical Problems

A comparison between the one-parametrical problem (3.30) and the optimization problem (3.19) shows, that the conditions differ only in the constraints on the surface S_p. In the optimization problem, these are the component-wise constraints on the displacement velocities. This may be clearly seen, if we take into account that:

$$\int\limits_{V} \lambda^+ \dfrac{\partial f(\sigma^- + \rho)}{\partial P^+} \, dv + \int\limits_{V} \lambda^- \dfrac{\partial f(-\sigma^- - \rho)}{\partial P^+} \, dv = \dot{u}^+,$$

$$(3.31)$$

$$\int\limits_{V} \lambda^+ \dfrac{\partial f(\sigma^+ + \rho)}{\partial P^-} \, dv + \int\limits_{V} \lambda^- \dfrac{\partial f(-\sigma^- - \rho)}{\partial P^-} \, dv = \dot{u}^-$$

Then we have:

$$\dot{u}^+ \geq T^+, \quad \dot{u}^- \geq T^- \qquad (3.32)$$

Hence, the weight multipliers of the optimality criterion, as in case of monotonically increasing loading, may be related to strains expressing the lower bound of the displacement velocities.

In the one-parametrical problem the integral form of the constraints on he displacement velocities (3.29) may be looked upon as the fixed power, if both terms of (3.29) are multiplied by $P°$. Thus in the optimization problem, the constraints on the displacement velocities may be selected more freely. Thus, in the optimization problem a more free selection of the constraints on the displacement velocities enables the optimal cyclic loading to be obtained. Besides, every component can be evaluated separately. It is obvious that in a problem of this kind we cannot obtain one factor of safety for the entire loading, as in the one-parametrical problem. The factor of safety can be determined only for every separate loading component. If the load factor has to be changed, then the value of the weight multiplier of the optimality criterion will be changed respectively.

3.3. Body Design Problem

In the body design problem, the vector fields of the extremum loads are prescribed completely by their value and direction. The purpose of the present problem is to determine the body composition, i.e. the distribution of the plasticity constant satisfying the optimality criterion (1.28).

The solution of the problem may depend on two different primal conditions:

1) the constant C of the material does not depend on the elastic constants of the material.
2) the constant C of the material depends on the elastic parameters of the body.

In the first case, the extremum "elastic" stresses σ^+ and σ^- do not depend on C and are assumed to be constant, when varying the yield function with respect to the variable C. The second case includes the problems, where the influence operators of the elastic solution ω^+ and ω^- depend on C. In this case, one of the possible ways of solving such problems is by iterating the solutions of some problems of the first type.

Hereinafter, the first case is dealt with.

3.3.1. Static Formulation

We derive the mathematical model from the minimum value of plastic dissipation. This theorem is written as the extremum problem:

$$(3.33) \qquad \min \int_V (\lambda^+ + \lambda^-) C \, dv$$

subject to

$$
\left.
\begin{array}{l}
\left.
\begin{array}{l}
C - f(\sigma^+ + \rho) \geqslant 0 \\
C - f(-\sigma^- - \rho) \geqslant 0 \\
C \geqslant 0 \\
\nabla\rho = 0
\end{array}
\right\} \ \text{in V,} \\
N\rho - r = 0 \ \ \text{on } S_u, \\
N\rho = 0 \ \ \text{on } S_p.
\end{array}
\right\}
\qquad (3.34)
$$

Since the scalar fields λ^+ and λ^- are not included in the conditions of the problem (3.34), the minimization of (3.33) is possible at any fixed values of them, including $\lambda^+ + \lambda^- \equiv \Lambda$ where Λ is the scalar field of the weight multipliers of the optimality criterion of the body. Thus, we obtain:

$$
\min \int_V \Lambda C \, dv
\qquad (3.35)
$$

subject to

$$
\left.
\begin{array}{l}
\left.
\begin{array}{l}
C - f(\sigma^+ + \rho) \geqslant 0 \\
C - f(-\sigma^- - \rho) \geqslant 0 \\
C \geqslant 0 \\
\nabla\rho = 0
\end{array}
\right\} \ \text{in V,} \\
N\rho - r = 0 \ \ \text{on } S_u, \\
N\rho = 0 \ \ \text{on } S_p.
\end{array}
\right\}
\qquad (3.36)
$$

This is the mathematical model of the body design problem for an elastic-plastic body under cyclic loading. The vector field of the residual stresses ρ and the scalar field of the plasticity constant C are sought. The problem is the functional analogue of a convex programming problem. Hereinafter, it is referred to as the static formulation of the body design problem.

3.3.2. Kinematic Formulation

The kinematic formulation of the problem is derived on the ground of the theory of duality. The Lagrange functional for the problem (2.32) - (2.33) has the form:

$$F(C,\rho,r,\lambda^+,\lambda^-,\nu,\dot{w}) = \int_V \Lambda C\, dv - \int_V \lambda^+[C - f(\sigma^+ + \rho)]\, dv -$$

(3.37)
$$- \int_V \lambda^-[C - f(-\sigma^- - \rho)]\, dv - \int_V \nu C\, dv - \int_V \dot{w}^T \cdot \nabla\rho\, dv -$$

$$- \int_{S_u} \dot{w}^T \cdot (N\rho - r)\, ds - \int_{S_p} \dot{w}^T \cdot N\rho\, ds .$$

Its variation with respect to the variables of the primal problem (C,ρ,r) when set equal to zero, yield the condition of the dual problem:

(3.38)
$$\left.\begin{array}{l} \Lambda - \lambda^+ - \lambda^- - \nu = 0 \\[4pt] \lambda^+ \geqslant 0, \quad \lambda^- \geqslant 0, \ \nu \geqslant 0 \\[10pt] \lambda^+ \dfrac{\partial f(\sigma^+ + \rho)}{\partial\rho} + \lambda^- \dfrac{\partial f(-\sigma^- - \rho)}{\partial\rho} - \nabla^T \dot{w} = 0 \\[10pt] \dot{w} = 0 \quad \text{on } S_u . \end{array}\right\} \text{ in V,}$$

Substituting (3.38) in (3.37), and taking into account that the plastic dissipation at the residual velocities is zero, we obtain the kinematic formulation of the body design problem:

(3.39)
$$\max \left\{ \int_V \lambda^+ f(\sigma^+ + \rho)\, dv + \int_V \lambda^- f(-\sigma^- - \rho)\, dv \right\}$$

subject to

$$
\left.
\begin{array}{l}
\left.
\begin{array}{l}
\lambda^{+} \dfrac{\partial f(\sigma^{+}+\rho)}{\partial \rho} + \lambda^{-} \dfrac{\partial f(-\sigma^{-}-\rho)}{\partial \rho} - \nabla^{T} \dot{w} = 0 \\[4mm]
\lambda^{+} \geqslant 0, \quad \lambda^{-} \geqslant 0 \\[3mm]
\lambda^{+} + \lambda^{-} \leqslant \Lambda
\end{array}
\right\} \text{ in V,} \\[10mm]
\dot{w} = 0 \quad \text{on } S_{u}.
\end{array}
\right\} \quad (3.40)
$$

where the vector fields ρ and $\dot{w}$ and the scalar fields λ^{+} and λ^{-} are sought.

Let us explain the physical sense of the problem. The first and the last conditions in (3.40) express the kinematically possible field of the velocities of the residual displacements. The inequalities express the concept of admissibility of the field. These are a form of normalization and enable the field of velocities of residual displacements to be derived to within a constant multiplier. Thus, all conditions (3.40) taken together express the kinematically admissible field of velocities of residual displacements. The objective function expresses power per cycle. This statement is proved by the relations which hold for a constant multiplier:

$$
\int_{V} \lambda^{+} f(\sigma^{+}+\rho)\,dv + \int_{V} \lambda^{-} f(-\sigma^{-}-\rho)\,dv =
$$

$$
= \int_{V}\left[\lambda^{+} \frac{\partial f(\sigma^{+}+\rho)}{\partial \rho}\right]^{T} \cdot \sigma^{+}dv + \int_{V}\left[\lambda^{-} \frac{\partial f(-\sigma^{-}-\rho)}{\partial \rho}\right]^{T} \cdot \sigma^{-}dv =
$$

$$
= \iint_{V\,S_{p}}\left[\lambda^{+} \frac{\partial f(\sigma^{+}+\rho)}{\partial \rho}\right]^{T} \cdot [\omega^{+} P^{+} - \omega^{-} P^{-}]\,dvds +
$$

$$
+ \iint_{V\,S_{p}}\left[\lambda^{-} \frac{\partial f(-\sigma^{-}-\rho)}{\partial \rho}\right]^{T} \cdot [\omega^{-} P^{+} - \omega^{+} P^{-}]\,dvds = \qquad (3.41)
$$

$$
= \iint_{V\,S_{p}}\left[\omega^{+T}\lambda^{+} \frac{\partial f(\sigma^{+}+\rho)}{\partial \rho} + \omega^{-T}\lambda^{-} \frac{\partial f(-\sigma^{-}-\rho)}{\partial \rho}\right]^{T} \cdot P^{+}\,dvds -
$$

$$-\iint_{V\,S_p} \left[\omega^{-T}\lambda + \frac{\partial f(\sigma^+ + \rho)}{\partial \rho} + \omega^{+T}\lambda - \frac{\partial f(-\sigma^- - \rho)}{\partial \rho} \right] \cdot P^- \, dvds \;\; =$$

$$= \int_V \int_{S_p} [\omega^{+T}\dot{\epsilon}^+ - \omega^{-T}\dot{\epsilon}^-]^T \cdot P^+ \, dvds \; - \iint_{V\,S_p} [\omega^{-T}\dot{\epsilon}^+ +$$

$$+ \omega^{+T}\dot{\epsilon}^-]^T \cdot P^- \, dvds = \int_{S_p} \dot{u}^{+T} \cdot P^+ \, ds + \int_{S_p} \dot{u}^{-T} \cdot P^- \, ds .$$

(3.41)

The mathematical model (3.39)   (3.40), therefore, corresponds to the extremum theorem:

Of all kinematically admissible fields of velocities of residual displacements at cyclic plastic failure, that one is actual which corresponds to the maximum value of external power in a cycle.

Hereinafter, this theorem will be referred to as the kinematic theorem of cyclic plastic failure. It is dual to the static theorem.

3.3.3. Theorems of Duality

The first theorems of duality for the pair of problems (3.35) - (3.36) and (3.39) (3.40) will be:

$$(3.42) \qquad \int_V \Lambda C^* \, dv = \int_V \lambda^{+*} f(\sigma^+ + \rho^*) \, dv + \int_V \lambda^{-*} f(-\sigma^- - \rho^*) \, dv .$$

From a physical point of view, this implies that the plastic dissipation and the external power in a cycle of statically and kinematically admissible fields of residual stresses and residual displacement velocities (deformations) are equal.

The second theorem of duality yields the associated cyclic plastic flow rule having the form (3.17) or (3.18).

The generalized Lagrange problem consists of all constraints and additional orthogonality conditions:

$$c - f(\sigma^+ + \rho) \geqslant 0$$

$$c - f(-\sigma^- - \rho) \geqslant 0$$

$$c \geqslant 0$$

$$\nabla \rho = 0$$

$$\lambda^+ \frac{\partial f(\sigma^+ + \rho)}{\partial \rho} + \lambda^- \frac{\partial f(-\sigma^- - \rho)}{\partial \rho} - \nabla^T \dot{w} = 0$$

$$\lambda^+ \geqslant 0, \quad \lambda^- \geqslant 0$$

$$\lambda^+ + \lambda^- \leqslant \Lambda$$

$$\lambda^+ > 0, \text{ subject to } c - f(\sigma^+ + \rho) = 0$$

$$\lambda^+ = 0, \text{ subject to } c - f(\sigma^+ + \rho) > 0$$

$$\lambda^- > 0, \text{ subject to } c - f(-\sigma^- - \rho) = 0$$

$$\lambda^- = 0, \text{ subject to } c - f(-\sigma^- - \rho) > 0$$

$$\sigma^+ = \int_{S_p} (\omega^+ P^+ - \omega^- P^-) ds,$$

$$\sigma^- = \int_{S_p} (\omega^- P^+ - \omega^+ P^-) ds$$

$$P^+ \geqslant 0, \quad P^- \geqslant 0$$

$$N\rho - r = 0, \quad \dot{w} = 0 \text{ on } S_u,$$

$$N\rho = 0 \quad \text{on } S_p.$$

in V,

$$(3.43)$$

3.3.4. Proof of Static Theorem Cyclic Plastic Failure

As in the case of the load design problem, the Kuhn-Tucker relations will be used to prove the primal theorem leading to the mathematical models of the body problem. The field of the residual stresses is denoted by an asterisk. It satisfies the conditions of the statically admissible field (3.36) and corresponds to the minimum value of the plastic dissipation (3.35).

Let there exist some actual field ρ satisfying the conditions (3.36), and only one distribution of the plasticity constant C for the body. Then the Kuhn-Tucker relations for the Lagrange functional (3.37) will have the form:

(3.44)
$$F(C,\rho,r,\lambda^{+*},\lambda^{-*},\nu^*,\ddot{w}^*) -$$
$$- F(C^*,\rho^*,r^*,\lambda^{+*},\lambda^{-*},\nu^*,\ddot{w}^*) \geqslant 0.$$

Substituting the Lagrange functional (3.37) in (3.34), we obtain:

(3.45)
$$\int_V \lambda^{+*}[\,f(\sigma^+ + \rho) - f(\sigma^+ + \rho^*)\,]\,dv \; +$$
$$+ \int_V \lambda^{-*}[\,f(-\sigma^- - \rho) - f(-\sigma^- - \rho^*)\,]\,dv \geqslant 0.$$

Since for the optimal solution $\lambda^{+*} f(\sigma^+ + \rho^*) = \lambda^{+*} C$ and $\lambda^{-*} f(-\sigma^- - \rho^*) = \lambda^{-*} C$ the difference in the square brackets in (3.45), as the yield conditions imply, cannot be positive.

Hence, the expression (3.45) is satisfied only as equality. Then every addend is equal to zero, i.e. we obtain:

$$f(\sigma^+ + \rho) = f(\sigma^+ + \rho^*),$$
$$f(-\sigma^- - \rho) = f(-\sigma^- - \rho^*).$$

It follows that:

$$\rho = \rho^*$$

Thus, of all fields of residual stresses at cyclic plastic failure the actual one corresponds to the minimum value of plastic dissipation. The theorem has been proved.

3.3.5. One-Parametrical Problem

The mathematical models for the one-parametrical body design problem will be derived.

Let the distribution of the plasticity constant of the material in the volume V be prescribed by:

$$C = C^\circ \gamma \qquad (3.46)$$

where C° is the required parameter and γ is the prescribed scalar field of the distribution of the plasticity constant. Then the plastic dissipation is expressed by:

$$D = C^\circ \int_V (\lambda^+ + \lambda^-) \gamma dv. \qquad (3.47)$$

Fixing the multiplier $\int_V (\lambda^+ + \lambda^-) \gamma dv \equiv 1$ we have a dual pair of problems determining the parameter C° of the plasticity constant under cyclic loading:

a) **Static Formulation**

$$\min C^\circ \qquad (3.48)$$

subject to

$$\left.\begin{array}{r} C^\circ\gamma - f(\sigma^+ + \rho) \geqslant 0 \\ C^\circ\gamma - f(-\sigma^- - \rho) \geqslant 0 \\ C^\circ \geqslant 0 \\ \nabla\rho = 0 \\ \\ N\rho - r = 0 \quad \text{on } S_u , \\ N\rho = 0 \quad \text{on } S_p ; \end{array}\right\} \text{in V,} \qquad (3.49)$$

b) **Kinematic Formulation**

$$\max \left\{ \int_V \lambda^+ f(\sigma^+ + \rho) dv + \int_V \lambda^- f(-\sigma^- - \rho) dv \right\} \qquad (3.50)$$

subject to

$$(3.51) \quad \left. \begin{array}{l} \lambda^+ \dfrac{\partial f(\sigma^+ + \rho)}{\partial \rho} + \lambda^- \dfrac{\partial f(-\sigma^- - \rho)}{\partial \rho} - \nabla^T \dot{w} = 0 \\[2ex] \lambda^+ \geqslant 0, \quad \lambda^- \geqslant 0 \\[1ex] \int_V (\lambda^+ + \lambda^-)\gamma dv \leqslant 1 \\[2ex] \dot{w} = 0 \text{ on } S . \end{array} \right\} \text{ in } V,$$

The theorems of duality hold for this pair of problems. According to the first theorem, we have:

$$(3.52) \quad C^{o*} = \int_V \lambda^{+*} f(\sigma^+ + \rho^*) dv + \int_V \lambda^{-*} f(-\sigma^- - \rho^*) dv .$$

From a physical point of view, this expression means that the parameter of the plasticity constant is equal to the external power for the actual field of the velocities of the residual displacements.

The generalized Lagrange problem for the dual pair of one-parametrical problems has the form:

$$(3.53) \quad \left. \begin{array}{l} C^o \gamma - f(\sigma + \rho) \geqslant 0 \\[1ex] C^o \gamma - f(-\sigma - \rho) \geqslant 0 \\[1ex] C^o \geqslant 0 \\[1ex] \nabla \rho = 0 \\[3ex] \lambda^+ \dfrac{\partial f(\sigma^+ + \rho)}{\partial \rho} + \lambda^- \dfrac{\partial f(-\sigma^- - \rho)}{\partial \rho} - \nabla^T \dot{w} = 0 \\[2ex] \lambda^+ \geqslant 0, \ \lambda^- \geqslant 0 \\[1ex] \int_V (\lambda^+ + \lambda^-)\gamma dv \leqslant 1 \end{array} \right\} \text{ in } V,$$

$$\lambda^+ > 0, \quad \text{subject to} \quad C^\circ\gamma - f(\sigma^+ + \rho) = 0$$
$$\lambda^+ = 0, \quad \text{subject to} \quad C^\circ\gamma - f(\sigma^+ + \rho) > 0$$
$$\lambda^- > 0, \quad \text{subject to} \quad C^\circ\gamma - f(-\sigma^- - \rho) = 0$$
$$\lambda^- = 0, \quad \text{subject to} \quad C^\circ\gamma - f(-\sigma^- - \rho) > 0 \quad \Big\} \quad \text{in V},$$

$$\sigma^+ = \int_{S_p} (\omega^+ P^+ - \omega^- P^-)ds$$
$$\sigma^- = \int_{S_p} (\omega^- P^+ - \omega^+ P^-)ds \qquad (3.53)$$
$$P^+ \geqslant 0, \quad P^- \geqslant 0$$
$$N\rho - r = 0, \quad w = 0 \quad \text{on } S_u,$$
$$N\rho = 0 \quad \text{on } S_p.$$

The fields $\rho, r, \lambda^+, \lambda^-, w$ and the parameter C° satisfying the conditions (3.53), are the solution of the problem determining the parameter of the plasticity constant under cyclic loading.

3.3.6. Comparison of Conditions of Optimization and One-Parametrical Problems

Comparing the conditions of the one-parametrical body design problem (3.53) and those of the optimization problem (3.43), we see that they differ in the constraints on the scalar fields of the multipliers λ^+ and λ^-. The optimization problem provides the component-wise constraints $\lambda^+ + \lambda^- \leqslant \Lambda$ while in the one-parametrical problem they are given in the integral form $\int_V (\lambda^+ + \lambda^-) \times \gamma dv \leqslant 0$. Both conditions are of normalization character and enable the fields of the residual displacement velocities to be obtained to within a constant multiplier. From a physical point of view, these constraints may be regarded as the fixed plastic dissipation in cycle. This becomes evident, if we multiply the constraints $\lambda^+ + \lambda^- \leqslant \Lambda$ by C and then integrate them in V, and if the latter are multiplied by C°. Then we have:

(3.54) $$\int\limits_V (\lambda^+ + \lambda^-) C\, dv \leqslant \int\limits_V \Lambda C\, dv$$

and

(3.55) $$C^\circ \int\limits_V (\lambda^+ + \lambda^-) \gamma dv \leqslant C^\circ \cdot 1.$$

This transformation is somewhat artificial, and besides, the condition $\lambda^+ + \lambda^- \leqslant \Lambda$ gives tighter constraints than (3.54). In addition, it permits a free selection of the constraints and enables the optimal distribution of the plasticity constant to be obtained.

3.4. Comparison of Mathematical Models of Monotonically Increasing and Cyclic Loading

The second chapter of the present paper was concerned with the derivation of mathematical models for monotonically increasing loading, cyclic loading being dealt with in the third chapter. The major difference lies in the values to be found. The main unknowns in the mathematical models for monotonically increasing loading are the fields of actual stresses and displacement velocities, while those in the mathematical models for cyclic loading are the fields of residual stresses and displacement velocities. The problems have in common, that the actual fields correspond to the maximum value of external power and to the minimum value of plastic dissipation. The mathematical models and correspondingly the theorems for monotonically increasing loading will be shown to be a particular case of the mathematical models and theorems for cyclic loading. The primal static formulations will be proved. As mentioned above, the kinematic formulation need not be proved as these are obtained formally throughout the theory of duality.

3.4.1. Load Design Problem

The static formulation the mathematical model for cyclic loading has the form (3.7) - (3.8). Let us consider this model for monotonically increasing loading. The process of monotonically increasing loading may be presented as a cycle determined by one specific value, i.e. the field $P = P^+ \equiv -P^-$ corresponding to a single distribution of "elastic" stresses. We denote this field by σ° i.e. $\sigma^+ = -\sigma^- = \sigma^\circ = \int\limits_{S_p} \omega P\, ds$. Now the vector field of the weight multipliers of optimality criterion has the form

$T^+ + T^- = T$. Substituting there expressions in (3.7) - (3.8), we obtain:

$$\max \int_{S_p} T^T \cdot P e \, ds \qquad (3.56)$$

subject to

$$\left. \begin{array}{r} f(\sigma^\circ + \rho) \leqslant C \\ \nabla\rho = 0 \end{array} \right\} \text{ in } V,$$

$$N\rho - r = 0 \quad \text{on } S_u,$$

$$\left. \begin{array}{r} N\rho = 0 \\ P \geqslant 0 \end{array} \right\} \quad \text{on } S_p . \qquad (3.57)$$

Let us change the variables by introducing the actual stresses $\sigma = \sigma^\circ + \rho$ and the response $R = N\sigma^\circ + r$. Substituting these expressions in (3.57) and taking into account that the "elastic" stresses are statically compatible, i.e. $N\sigma^\circ = Pe$ we have:

$$\max \int_{S_p} T^T \cdot P e \, ds$$

subject to

$$\left. \begin{array}{r} f(\sigma) \leqslant C \\ \nabla\sigma = 0 \end{array} \right\} \text{ in } V,$$

$$N\sigma - R = 0 \quad \text{on } S_u,$$

$$\left. \begin{array}{r} N\sigma - Pe = 0 \\ P \geqslant 0 \end{array} \right\} \quad \text{on } S_p$$

The mathematical model obtained is similar to (2.3) - (2.4). Thus, it is obvious that the main theorems for a rigid-plastic body expressed by the stress and strain rates are only a particular case of the theorems for an elastic- plastic body expressed by the residual stresses and residual displacement velocities. Note, that in the conditions (3.54) the "elastic" stresses could be substituted by any field σ° satisfying only the static compatibility conditions. Thus, we obtain the field P corresponding to simple plastic failure, and the field ρ which does not express the actual residual stresses after the load has been removed.

Nevertheless, the sum of the statically compatible σ° field and the field ρ expresses the actual stresses at plastic failure in places where the strain rates are not equal to zero.

3.4.2. Body Design Problem

The calculations are carried out in the same way as in the load design problem. The mathematical model for monotonically increasing loading may be derived formally from the mathematical model (3.35) - (3.36). Using the definition of "elastic" stresses, we obtain:

(3.58)
$$\min \int_V \Lambda C \, dv$$

subject to

(3.59)
$$\left.\begin{array}{l} C - f(\sigma^\circ + \rho) \geqslant 0 \\ C \geqslant 0 \\ \nabla\rho = 0 \\ N\rho - r = 0 \quad \text{on } S_u, \\ N\rho = 0 \quad \text{on } S_p. \end{array} \left.\begin{array}{l} \\ \\ \end{array}\right\} \text{in } V, \right\}$$

Replacing the variables $(\sigma = \sigma^\circ + \rho, R = N\sigma^\circ + r)$, we obtain the mathematical model for monotonically increasing loading having the form (2.32) - (2.33).

APPENDIX

1. Formulation of Dual Problems of Convex Programming

Some definitions used in convex programming and the main results of the theory of duality will be discussed.

The purpose of mathematical programming problem is the derivation of the lower (upper) bound of a function (functional) subject to certain constraints. The qualitative analysis of the problem consists in the investigation of the type of the function and the required extremum. In addition, it is necessary to determine whether the set corresponding to the constraints is convex or not.

An arbitrary totality of the points of the n-dimensional space E^n will be referred to as a set. If the set G belongs to the space E^n, it is denoted by the relation $G \subset E^n$.

If x is an element of the set G, then $x \in G$. The set may be prescribed by a certain property of all its elements and only these elements. If, for example, G is a set of the points $x = (x_1, x_2 ... x_n)$, corresponding to the relation $\varphi(x) \leqslant b$ then we can write $G = \{x : \varphi(x) \leqslant b\}$. A set including all boundary points is referred to as a closed one. If a set does not include any boundary points (all points being inner), then it is referred to as open. For example $G = \{x : Ax \leqslant b\}$ is a closed set, and $G = \{x : Ax < b\}$ is an open set.

A convex set possesses the following property: if two points belong to the set, this conditions is:

$$x = \lambda x_1 + (1 - \lambda)x_2 \in G,$$

where

$$0 < \lambda < 1, \quad x_1, x_2 \in G.$$

A set having only one point is supposed to be convex. A non-negative octant, i.e. the set of points $x \geqslant 0$ is closed and convex. In order to find whether the set, whose elements correspond to the relations $\varphi(x) \leqslant b$ or $\varphi(x) \geqslant b$ is convex, the type of the function $\varphi(x)$ must be known. The function $\varphi(x)$ prescribed on the convex set $G \in E^n$ is referred to as a concave (convex), if the inequality:

$$\varphi[\lambda x_1 + (1 - \lambda)x_2] \leqslant (\geqslant)\lambda\varphi(x_1) + (1 - \lambda)\varphi(x_2). \tag{1}$$

holds for any points $x_1, x_2 \in G$ and for any value of λ, $0 < \lambda < 1$.

In other words, the function $\varphi(x)$ is concave, if the chord connecting any two points lies above it; the function is convex, if the chord lies below.

If the relation (1) has the sign of a strict inequality $<(>)$, then the function will be referred to as concave (convex), respectively. Note, that if   is a concave function, then multiplying it by -1, we shall obtain a convex function, and vice versa. Let us take some examples. The nonnegative quadratic form $\varphi(x) = x^T Dx$ is an essentially concave function throughout E^n; the linear form $\varphi(x) = Ax$ is concave and convex throughout E^n since for any two points x_1 and x_2 and for any λ we have:

$$A[\lambda x_1 + (1-\lambda)x_2] = \lambda Ax_1 + (1-\lambda)Ax_2.$$

Nevertheless, the function is neither essentially concave nor convex.

Now let us turn to the convexity criteria for the set, resulting from the concave and convex functions on the nonnegative octant E^n.

If the function $\varphi(x)$ is concave (convex), then the set satisfying the conditions $\varphi(x) \leq b$ and $x \geq 0$ ($\varphi(x) \geq b$ and $x \geq 0$), is convex.

Since the intersectional of convex sets is convex, then the set satisfying the conditions:

$$(2) \qquad \varphi_j(x)\{\leq, \geq\}b_j, \quad j = 1, 2, \ldots, m, \quad x \geq 0,$$

is convex, if functions φ_j are concave in the inequalities having the sign $\leq$ and if the functions φ_i are convex for $x \geq 0$ in the inequalities having the sign $\geq$.

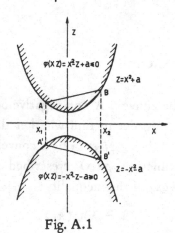

Fig. A.1

Thus, we have stated the criteria for the determination of convex sets. These statements are illustrated on the curves for the functions of one variable $z = x^2 + a$ $z = -x^2 - a$ in Fig. A.1. The inner domain of the parabola is convex in both cases, the outer one is non-convex.

Now we shall determine the conditions when any relative extremum of the concave (convex) function is its global extremum.

If $f(x)$ is a concave function prescribed on the closed convex set $G \in E^n$ then any relative minimum of $f(x)$ on G is the global minimum of $f(x)$ on G.

If $f(x)$ is a convex function prescribed on the closed convex set $G \subset E^n$ then any relative maximum of $f(x)$, on G is the global maximum of $f(x)$ on G.

The essentially concave (convex) function reaches its global minimum (maximum) at a single point.

Thus, in the analysis of extremum problems first is necessary to give a qualitative assessment of the type of the function (functional) whose extremum is sought, and then to investigate the set it is prescribed on, i.e. to determine whether the conditions of the problem are convex or not. The method of solving the problem may be chosen. The convex programming techniques makes possible the determination of the global minimum of concave functions and the global maximum of convex functions for sets (domains). Note, that for a linear function on the convex set this technique enables both the global minimum and the global maximum to be found.

Now, the main results of the theory of duality for convex programming problems will be dealt with.

Consider extremum problem:

$$\min f(x) \tag{3}$$

subject to

$$\left.\begin{array}{ll} \varphi_j(x) \leqslant b_j, & j = 1,2,\ldots,m_1 \leqslant m \\ \varphi_j(x) = b_j, & j = m_1 + 1, \ m_1 + 2,\ldots,m \\ x_i \geqslant 0, & i = 1,2,\ldots,n_1 \leqslant n \end{array}\right\} \tag{4}$$

where $x = (x_1, x_2, \ldots, x_n)^T$. The functions $f(x)$ and $\varphi_j(x)$ $j = 1,2,\ldots,m_1$ are concave, and the functions $\varphi_j(x)$, $j = m + 1, \ldots, m$ are linear. The first derivatives of the functions $f(x)$ and $\varphi_j(x)$ are continuous.

The set G determined by the conditions (4) is not empty, and according to (2) it is convex. Thus, (4) is a convex programming problem.

Hereinafter, the problem (3) - (4) will be referred to as primary. Its dual is based on the conditions:

1. The objective function of the dual problem is maximized instead of being minimized, as is the case in the primary problem.
2. the objective function of the dual problem is the Lagrange function:

$$(5) \qquad F(x,z) = f(x) - \sum_{j=1}^{m} z_j [b_j - \varphi_j (x)] ,$$

3. Every j condition of the primary problem corresponds to the new variable z of the dual problem. If the condition is expressed by an inequality, then the relevant variable is limited by the sign, i.e. it is not independent ($z_j \geqslant 0$); if the j condition is an equality, then the variable z_j is not limited by the sign, i.e. it is independent.
4. Every non-independent variable x_i of the primary problem corresponds to the condition in the dual problems expressed by the inequality:

$$(6) \qquad \frac{\partial f(x)}{\partial x_i} + \sum_{j=1}^{m} z_j \frac{\partial \varphi_j (x)}{\partial x_i} \geqslant 0, \quad i = 1,2,...,n_1$$

and every independent variable x_i corresponds to the equality:

$$(7) \qquad \frac{\partial f(x)}{\partial x_i} + \sum_{j=1}^{m} z_j \frac{\partial \varphi_j (x)}{\partial x_i} = 0, \quad i = n_1 + 1,...,n.$$

Thus, these statements lead to the pair of dual problems

$$\min f(x) \qquad \bigg| \qquad \max F(x,z) = f(x) - \sum_{j=1}^{m} z_j [b_j - \varphi_j (x)]$$

$$\varphi_j (x) \leqslant b_j ,$$

$$j = 1,2,...,m_1 \leqslant m \qquad \frac{\partial f(x)}{\partial x_i} + \sum_{j=1}^{m} z_j \frac{\partial \varphi_j (x)}{\partial x_i} \geqslant 0, \quad i = 1,2,...,n_1$$

$$\varphi_j (x) = b_j ,$$

$$j = m_1 + 1,...,m$$

$$x_i \geqslant 0, \qquad \frac{\partial f(x)}{\partial x_i} + \sum_{j=1}^{m} z_j \frac{\partial \varphi_j (x)}{\partial x_i} = 0, \quad i = n_1 + 1,...,n$$

$$i = 1,2,...,n_1 \leqslant n \qquad z_j \geqslant 0, \quad j = 1,2,...,m_1 .$$

$$(8)$$

The convex programming problem (3) - (4) is said to satisfy the Slaters condition if there exists any point $x°$ corresponding to the constraint (4), and if $\varphi_j(x°) < b_j$ $j = 1, 2, ..., m_1$.

The dual pair of problems (8) conforms to the statements:

1. If x^* is the solution of the primary problem, satisfying the Slater condition then we can find such T-dimensional vector that the vector (x^*, z^*) will be the solution of the dual problem, the extremum values of the objective functions of the dual problems being equal:

$$f(x^*) = F(x^*, z^*) . \qquad (9)$$

The equality (9) is usually referred to as the first theorem of duality.

2. If $[x^*, (x^*, z^*)]$ are the solution of dual problems (9). then the following relations hold true:

$$z_j^*[b_j - \varphi_j(x^*)] = 0, \quad j = 1, 2, ..., m \qquad (10)$$

This relation expresses the second theorem of duality (additional orthogonality condition).

3. In order for the vector x^* to be the solution of the primary problem, it is sufficient and necessary, provided the Slater condition is observed, to have such vector z^* that the vector (x^*, z^*) should satisfy the relations:

$$
\left.
\begin{aligned}
&\varphi_j(x^*) \leqslant b_j , \quad j = 1, 2, ..., m_1 \leqslant m, \\
&\varphi_j(x^*) = b_j , \quad j = m_1 + 1, ..., m, \\[2mm]
&\frac{\partial f(x^*)}{\partial x_i} + \sum_{j=1}^{m} z_j^* \frac{\partial \varphi_j(x^*)}{\partial x_i} \geqslant 0, \quad i = 1, 2, ..., n_1 , \\[2mm]
&\frac{\partial f(x^*)}{\partial x_i} + \sum_{j=1}^{m} z_j^* \frac{\partial \varphi_j(x^*)}{\partial x_i} = 0, \quad i = n_1 + 1, ..., n, \\[2mm]
&x_i^* \geqslant 0, \quad i = 1, 2, ..., n_1 , \\
&z_j^* \geqslant 0, \quad j = 1, 2, ..., m_1 , \\
&z_j^*[b_j - \varphi_j(x^*)] = 0, \quad j = 1, 2, ..., m.
\end{aligned}
\right\}
\qquad (11)
$$

Now the sense of the aforementioned statement will be explained. The Lagrange function of the problem (3) is introduced:

$$F(x,z) = f(x) - \sum_{j=1}^{m} z_j [b_j - \varphi_j (x)] \;,$$

In this case the conditions (11) are analogous to the relation:

(12)
$$F(x^* ,z) \leqslant F(x^* ,z^*) \leqslant F(x,z^*),$$
$$z_j \geqslant 0, \quad j = 1,2,\dots,m_1 , \quad x_i \geqslant 0, \quad i = 1,2,\dots,n_1$$

which means that (x^*,z^*) is the saddle point of the Lagrange functional in (3).

Thus, the Slater condition being observed, the point is the solution of the primary problem (3) only in case there is such a vector $z_j^* (z_j \geqslant 0, j=1,2,\dots,m_1)$ for which (x^*,z^*) is the saddle point of the Lagrange function of the primary problem. This statement is referred to as the Kuhn-Tucker theorem. Though the above conclusions refer to the problems determined by the functions of the finite number of variables, these may be applied to functional analogous of convex programming problems, whose constraints are expressed, for example, by differential relations. These analogous are the limit design problems for solid deformable body. The conditions are the constraints including both functional and differential operators for the inner domains and boundary points of the body, while the extremum is determined for an integral type functional.

Nevertheless, the formulation of the problem, though more complicated, permit the theory of duality for convex programming problems to be applied.

NOTATIONS

The main notations of the geometrical and physical values used in the paper are given below. Vector fields are given in bold type and scalar ones - in standard type. To denote the velocities of the corresponding values the letters are marked with dots. The upper index "T" denotes transposition.

V	–	body volume
S	–	body surface
S_p	–	part of S; surface subjected to external actions (loading or displacements)
S_u	–	part of S; surface with zero displacements
x	–	coordinate of a particle of a body
t	–	time
τ	–	cycle period
c	–	plasticity constant of a material
$c°$	–	parameter of a plasticity constant
γ	–	distribution of a plasticity constant
f	–	yield function
P	–	loading on S_p
P^+	–	upper bound of loading
P^-	–	lower bound of loading
$P°$	–	loading parameter
η	–	distribution of external actions on S_p
η^+	–	distribution of upper bound of external actions
η^-	–	distribution of lower bound of external actions
e	–	direction of external actions on S_p
σ	–	stresses
σ^+	–	upper bound of extreme stresses
σ^-	–	lower bound of extreme stresses
$\sigma°$	–	"elastic" stresses
ρ	–	residual stresses
ϵ	–	deformations
u	–	displacements
λ	–	multiplier
λ^+	–	upper bound of multipliers
λ^-	–	lower bound of multipliers

∇ – differential operator of static equilibrium

N – algebraic operator of static equilibrium

ω – influence matrix of "elastic" solution for loading given on S_p

T – weight multiplier of optimality criterion for external actions

T^+ – upper bound of weight multipliers of optimality criterions

T^- – lower bound of weight multipliers of optimality criterions

Λ – weight multiplier of optimality criterion of a body

THEORY OF OPTIMAL
PLASTIC DESIGN
OF STRUCTURES

by

M.A. SAVE
Faculté Polytechnique de Mons
Belgium

CHAPTER I

OPTIMALITY CONDITION: THE SIMPLEST FORM

1.1. Plastic Analysis and Design

Consider a structure, the general layout of which is given (axes for systems of beams, midsurfaces for shells) together with the boundary conditions. Denote by ξ an arbitrary point on that layout, and by   the corresponding line or surface element. Let q_i , $\dot{q}_i$ and Q_i ($i = 1, ..., n$)   ie generalized strains, strain rates, and stresses that describe the behaviour of the structure, Q_i and $\dot{q}_i$ being chosen in such a manner that $\sum_i Q_i \dot{q}_i$ is the power D of the stresses Q_i on the strain rates $\dot{q}_i$,summation being extended over the range i = 1, ..., n.

Assume now standard perfect plasticity of the structure. This assumption means that:

a.   there exists, for each structural element, a convex yield locus with equation

$$F(Q_i) = 1 \qquad\qquad (1)$$

This locus does not depend on previous plastic deformations: it is fixed in the stress-space. For homogeneous structures it does not depend on ξ . All possible stress states satisfy

$$F(Q_i) \leqslant 1 \qquad\qquad (2)$$

Hence all stress vectors $\vec{Q}$ with components Q_i have their extremities inside or on the yield locus.

b.   for any stress point Q_i on the yield locus, local plastic flow is possible; the corresponding strain-rate vector $\vec{q}$ with components $\dot{q}_i$ is outward normal to the yield locus at that stress-point, when the strain-rate space q_i is superimposed on the stress-space Q_i . The magnitude of $\vec{q}$ is left undetermined (but non-negative).

We restrict our attention for the time being to one-parameter loading systems, that we describe in the following manner*

*) Concentrated loads will be obtained as uniformly distributed loads acting on very small areas.

(3) $$\vec{P}(\xi) = P \vec{a}(\xi)$$

where $\vec{a}(\xi)$ is a given field of vectors and P the scalar loading parameter.

Any stress field $\vec{Q}(\xi)$ satisfying the yield condition (2) and the equilibrium equations of the structure for some value of P will be called statically admissible, together with the corresponding value of the load parameter, that will be denoted by $\underline{P}$.

When enough plasticity has spread out in the structure upon increasing P, control of displacements may be lost, with the occurance of a plastic collapse mechanism. At incipient collapse under constant loads (all changes of geometry being neglected), the power of the loads is dissipated in heat is the plastic regions. If $\vec{v}(\xi)$ is the velocity field describing the collapse mechanism, $\vec{q}(\xi)$ the strain-rate field derived by kinematics from $\vec{v}(\xi)$, we thus have *

(4) $$\int \vec{P}(\xi) \cdot \vec{v}(\xi) d\xi = \int D(\xi) d\xi$$

From the normality law, the specific ** power of dissipation D is given by

(5) $$D = \vec{Q} \cdot \vec{q} = \sum_i Q_i \dot{q}_i$$

It can easily be shown that D is a single-valued function of $\vec{q}$. Using (3), (4) is re-written

(5) $$P \int \vec{a}(\xi) \vec{v}(\xi) d\xi = \int D[\vec{q}(\xi)] d\xi .$$

---

*) Except when otherwise indicated, integration is extended over the layout of the structure.
**) Specific means here per unit of length of axis or per unit area of mid-surface, that is per element of layout.

Defining

$$\mathscr{H} = \int \vec{a}(\xi)\,\vec{v}(\xi)\,d\xi \ , \\[4pt]
(6) \qquad \mathscr{P} = P\mathscr{H} \ , \\[4pt]
\mathscr{D} = \int D\,d\xi \ , \ \Bigg\}$$

the power equation (5) can be written as

$$P\mathscr{H} = \mathscr{D} \qquad \text{or} \qquad (7)$$

$$\mathscr{P} = \mathscr{D} \qquad\qquad (8)$$

Note that $\mathscr{D}$ is essentially positive, and so is $\mathscr{P}$. Various possible collapse mechanisms, constructed as described above (with positive $\mathscr{P}$) can be imagined. They are called kinematically admissible mechanisms, together with the corresponding load parameters given by equation 7 and denoted P_+.

The basic theorems of plastic limit analysis are then summarized in the following continued inequality:

$$P_- \leqslant P_\varrho \leqslant P_+ \ , \qquad (9)$$

where P_ϱ is the (real) collapse load parameter*.

In a design problem, loading $\vec{a}(\xi)$ is given together with a minimum collapse load P_ϱ (the structure must be designed in such a way that the real collapse load be at least equal to the assigned P_ϱ). Assuming that, for economy, the design collapses strictly at assigned P_ϱ we have, for any kinematically admissible mechanism, from (7),

$$P_+\mathscr{H} = \mathscr{D} \ , \quad \text{with} \quad P_+ \geqslant P_\varrho \ , \qquad (10)$$

From (10) we immediately obtain

$$P_\varrho \mathscr{H} \leqslant \mathscr{D} \qquad (11)$$

*) For the sake of brevity, we shall from now on simply say "load" for "load parameter".

or, equivalently

(12) $\mathscr{P} \leqslant \mathscr{D}$

Relation (12), which is fundamental in optimal plastic design, can be phrased
as follows:
"for any structure able to support the assigned collapse loads, the power of the loads
in any kinematically admissible mechanism cannot be larger than the plastic
dissipation".

1.2. Optimality Condition

We first consider the following simple plastic design problem:
  a) the design is defined by one unknown function t(ξ). The meaning of t will
     be discussed later.
  b) assigned loads are fixed in space, and consist of only one system of forces;
     they are independent of t(ξ).
  c) by definition, an optimal design will minimize the expression $\int t(\xi)d\xi$,
     called total cost $\mathscr{C}$ whereas t(ξ) is called specific cost.

Assume now that a first design t_1 is at collapse (under the assigned load P), with a
collapse mechanism described by $\vec{v}(\xi)$ and the related $\vec{q}(\xi)$. From equation (7)
we have

(13) $P\mathscr{H} = \mathscr{D}_1$

Consider any second design t_2 able to support at least P (or larger loads).
Regarding $\vec{q}(\xi)$ and $\vec{v}(\xi)$ as a kinematically admissible mechanism for design t_2,
and noting from (6) that $\mathscr{H}$ does not depend on t, we have, according to relation
(12),

(14) $P\mathscr{H} \leqslant \mathscr{D}_2$

Comparing (13) and (14) we conclude that

$$\mathscr{D}_1 \leqslant \mathscr{D}_2 \ ,$$

or, from the definition (6) of $\mathscr{D}$, that

$$\int D_1 \, d\xi \;\leqslant\; \int D_2 \, d\xi \;.$$

Hence, it is sufficient to satisfy the condition

$$\frac{D_1}{t_1} = \frac{D_2}{t_2} = \alpha > 0, \quad (15), \text{(with } \alpha \text{ independent of } \xi)$$

to obtain

$$\int t_1 \, d\xi \;\leqslant\; \int t_2 \, d\xi \qquad \text{and conclude that}$$

t_1 is a design of absolute optimality.

CHAPTER II

APPLICATION OF THE OPTIMALITY CONDITION AND DISCUSSION

2.1. Beam Example

To illustrate the significance and the application of the optimality condition (15), we shall treat in detail the beam example of Fig. 1. The cross-section of that beam is rectangular, with fixed height h and unknown variable breadth $b(\xi)$, which plays here the role of t in the theory of chapter 1. Because plane sections are assumed to remain plane and normal to the deformed axis, and with no axial loading, the only generalized strain rate is the rate of curvature $\dot{K}(\xi)$. The corresponding stress is the bending moment M, with value M_p at full plasticity of the cross-section.

Hence, the specific power of dissipation is

$$(16) \qquad D = M_p |\dot{K}| \quad ,$$

because the flow law is here simply

$$(17) \begin{cases} |M| = M_p \\ \text{sign } \dot{K} = \text{sign } M \, . \end{cases}$$

We also know that

$$(18) \qquad M_p = \frac{bh^2}{4} \sigma_y$$

where σ_y is the yield stress in pure tension or compression. Substitution of (18) in (16) gives

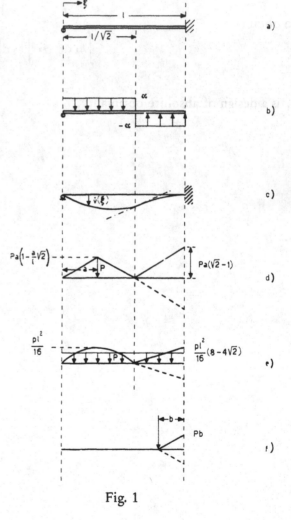

Fig. 1

$$\frac{D}{b} = \frac{h^2}{4} \sigma_y |\dot{K}| \tag{19}$$

Equation (19) shows us a fundamental feature of that type of structure, namely that the function D/b (and, in general D/t) is independant of the design variable. Hence, as $h^2/4\,\sigma_y > 0$, condition (15) reduces to

$$|\dot{K}| = \alpha > 0 , \tag{20}$$

a purely kinematical condition.

For all downward transversal loadings, condition (20) will enforce positive constant curvature in some region adjacent to the left-end simple support and constant negative curvature (with same magnitude) in the vicinity of the right, built-in end. Equilibrium of the conjugate beam of Fig. 1b furnishes the abscissa $\xi_0 = \ell/\sqrt{2}$ of the point of contraflexure. Fig.1c shows the collapse mechanisms $\dot{v}(\xi)$

After this kinematical part of the solution, expressing the optimality condition, the design itself is obtained from statics. Indeed, for the mechanisms of Fig. 1c to occur, the flow law (17) indicates that:

$$\left. \begin{array}{ll} \text{for } 0 \leqslant \xi \leqslant \ell/\sqrt{2}, & M_p = M \\[2mm] \ell/\sqrt{2} \leqslant \xi \leqslant \ell, & M_p = -M \end{array} \right\} \equiv M_p = |M| .$$

Continuity of $M(\xi)$ enforces $M = 0$ at $\xi = \ell/\sqrt{2}$, a condition that makes the beam statically determinate. For any load system of downward acting transversal forces, M is thus readily obtained, and also the design $b(\xi) = 4|M|/h^2\sigma_y$.

Examples of such designs are given in Fig. 1d, 1e and 1f, where M and $|M|$ are represented.

We can now make the following remarks:

a. we were able to apply the optimality condition (15) because D/t does not depend on t; because t is the only variable cross-sectional dimension, it is seen that optimality is reached through a collapse mechanism with constant dissipation per unit volume. This is the DRUCKER-SHIELD condition [1].

As will be seen in more detail later, cases where D/t is independent of t are:

— sandwich structures, made of thin perfectly plastic face-sheets that resist in-plane forces, and of a core without bending or membrane strength but infinitely resistant to transversal forces.

- disks in plane stress, with unknown variable (small) thickness.
- beams where the volume V per unit of length is related to the yield moment M_p by

$$V = a + b\,M_p \quad , \text{ where } \quad a \geqslant 0 \;, \; b > 0.$$

Indeed with $M_p(\,\xi\,)$ taken as the unknown design function, $D/M_p = |\dot{K}|$ is independent of M_p, and minimizing $\int V d\xi$ is equivalent to minimizing $\int M_p\, d\xi$.

It is worth noting that, if a were negative, as M_p and V are essentially non- negative, M_p would be forced not to decrease below -a/b, a situation that renders the problem quite different, as will be seen later.

b. the same collapse mechanism (with constant dissipation) is valid for a whole family of loadings, to which a superposition theorem holds, stating that, upon superposition of loads corresponding to the same collapse mechanism, the optimal design is obtained by simple addition of the optimal design for the separable loads. The proof of this theorem is left to the reader, as it is a rather direct consequence of the linearity of the equilibrium equations and of proportionality of M_p to the design variable. This theorem will be generalized in section 3.5 to sandwich shells.

2.2. Circular Plate Example

The circular plate, axisymmetrically loaded, is the simplest example that however retains all basic features of a general shell problem.

Consider a simply supported circular sandwich plate of a Tresca material, uniformly loaded. In circular coordinates r and θ (Fig. 2), the principal rates of curvature $\dot{K}_r$ and $\dot{K}_\theta$ are the only generalized strains, to which correspond the principal bending moments M_r and M_θ.

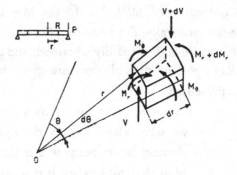

Fig. 2

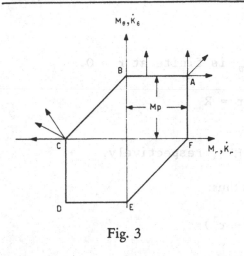

Fig. 3

The yield locus is the wellknown Tresca hexagon of Fig. 3,

where $\qquad M_p = \sigma_y Ht$, $\qquad$ (21)

with σ_y : yield stress

H : given constant core thickness

t : unknown thickness of a face sheet.

The downward transversal rate of displacement being denoted by $\dot{v}$, we know that

$$\dot{K}_r = - \frac{d^2 \dot{v}}{dr^2} \quad , \quad \dot{K}_\theta = - \frac{1}{r} \frac{d\dot{v}}{dr} \quad .$$

(22)

The specific dissipation is

$$D = M_\theta \dot{K}_\theta + M_r \dot{K}_r$$

(23)

In formulating the constant dissipation optimality condition, we must consider the possible locations of the stress point (M_r , M_θ) on the yield hexagon. For example, the stress point cannot lie on the side AF for a finite range of r because the normality law requires $\dot{K}_\theta = 0$ and, in view of (22) it is impossible to obtain $D/t = \alpha > 0$. It can be shown that, for similar reasons, the optimality condition can be satisfied only for stress regimes represented by points A or D and C or F. Obviously, for the present problem point A is relevant. Substitution of M_p for M_θ and M_r in the expression (23) of D, and use of (21) and (22), give the following optimality condition

$$\frac{d^2 \dot{v}}{dr^2} + \frac{1}{r} \frac{d\dot{v}}{dr} = - \alpha \quad ,$$

(24)

when it is noted that α is a positive but otherwise arbitrary constant.

Integration of (24) results in

$$\dot{v} = - \frac{1}{4} \alpha r^2 + \frac{1}{2} \alpha C_1 \log \frac{C_1}{r} + C_2$$

(25)

where C_1 and C_2 are integration constants.

Boundary conditions are:

$$\frac{d\dot{v}}{dr} = 0 \quad \text{at } r = 0 \text{ because } \dot{K}_\theta \text{ is finite at } r = 0.$$

and

$$\dot{v} = 0 \quad \text{at } r = R$$

They give

$$C_1 = 0 \quad \text{and} \quad C_2 = +\frac{1}{4}\alpha R^2 , \text{ respectively.}$$

The constant dissipation collapse mechanism is thus:

$$(26) \qquad\qquad \dot{v} = \frac{1}{4}\alpha (R^2 - r^2)$$

The design will now be obtained from statics. The equilibrium equation is

$$(27) \qquad\qquad \frac{1}{r}\frac{d^2 (rM_r)}{dr^2} - \frac{1}{r}\frac{dM_\theta}{dr} = -p$$

Because regime A is valid throughout the plate, equation (27) becomes

$$(28) \qquad\qquad \frac{1}{r}\frac{d^2}{dr^2}(rM_p) - \frac{1}{r}\frac{dM_p}{dr} = -p$$

a solution of which is of the type $M_p = ar^2 + br + c$ (29). Substitution of (29) in (28) shows that b must vanish and $a = -p/4$. At $r = R$, $M_r = 0$ requires $M_p = 0$ and hence $c = -aR^2$. We finally obtain the optimal design

$$(30) \qquad\qquad M_p = \sigma_y Ht(r) = \frac{p}{4}(R^2 - r^2)$$

The total volume $\mathcal{V}$ of the face sheet is

$$(31) \qquad\qquad \mathcal{V} = \int_0^R 2\pi r \cdot 2t \cdot dr = \frac{\pi p R^4}{4\sigma_y H}$$

It shows a 25 % saving with respect to the constant thickness sandwich plate with same core material, boundary conditions and loading. Various other problems of circular sandwich plates have been treated in the literature: [2] to [9].

2.3. The Solid Plate: Optimality Condition

It may be argued that the solid plate (homogeneous through its thickness) is of greater practical interest than the sandwich plate. In the case of a solid plate with thickness t, we have

$$M_p = \sigma_y \frac{t^2}{4} . \tag{32}$$

For all stress regimes and particularly for corners A, D, C and F, the dissipation D is proportional to t^2, and hence D/t is proportional to t. The optimality condition (15) is not purely kinematical any more. Because it involves the unknown function $t(\xi)$, its application would necessitate comparison of a design no. 1 with all possible design no. 2, a task obviously impossible. Consequently, we are forced to restrict ourselves to the search for a relative optimality condition, by comparing design no. 1 with a neighboring design $t_1 + \delta t$. If we retain only first order terms, equation (14) becomes

$$P\mathcal{H} \leqslant \mathcal{D}_1 + \int \delta t . \frac{\partial}{\partial t} D_1 . d\xi \tag{14'}$$

Comparison of (14') with (13) shows that

$$\int \delta t \frac{\partial}{\partial t} D_1 d\xi \geqslant 0 .$$

Hence, a sufficient condition for relative optimality is

$$\frac{\partial}{\partial t} D = \alpha > 0 . \tag{15'}$$

Because D is proportional to t, condition (15') can be written as

$$\frac{D}{t} = \alpha > 0 \tag{15''}$$

It must be emphasized, as discussed by Mroz [10], that (15') or (15") are conditions of stationarity of the total cost. Minimum is obtained for some stress regimes only. Because D/t depends on t, the optimality condition is more delicate to apply and only very few applications exist. This fact can be understood also in the following way. Let us take $t^2(\xi)$ as design unknown function, in order to recover independence of D/t^2 with respect to the design unknown. The specific cost, that is

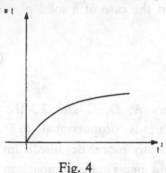

Fig. 4

the thickness t (volume per unit area) is then measured by the square root of the design variable. Its graph has the shape shown in Fig. 4. Clearly, the specific cost is a concave function of the design variable and our problem is one of concave optimization. Though they are generaly of great practical interest, the problems of concave optimization fall outside the scope of the present course and will thus be left aside.

Nevertheless, the solid plate example gives us the opportunity to stress the basic features of the class of problems we have chosen to treat from the very beginning:

— there is only one design unknown function $t(\xi)$
— the specific cost is identical (or proportional) to t
— structures treated are such that D/t is independent of t.

CHAPTER III

FIRST GENERALIZATION: MOVABLE LOADS

3.1. Optimality Condition

Suppose that the design must be optimized with the behavioral constraint consisting of not only one but some finite number of assigned systems of collapse loads, or more generally of one (or several) systems(s) of movable loads, the location of which, on the layout, depend on a* parameter λ varying in some range (λ_o , λ_1).

Consider a design t_1 at collapse for and only for some (or all) values of λ in the given range (λ_o , λ_1). Equation (13) can be written for all these values of λ. If t_2 is a design able to support the loads for all λ we can write equation (14) for every λ at which t_1 is at collapse, using the collapse mechanism of t_1 and obtain for each of these values of λ .

$$\mathscr{D}_1 \leqslant \mathscr{D}_2 \ , \ \text{or}$$

$$\int D_1 d\xi \ \leqslant \ \int D_2 d\xi \tag{32}$$

We can now integrate relation (32) over λ. After inverting the order of integration we obtain

$$\int \left[\int_{\lambda_0}^{\lambda_1} D_1 \, d\lambda \right] d\xi \ \leqslant \ \int \left[\int_{\lambda_0}^{\lambda_1} D_2 \, d\lambda \right] d\xi \tag{33}$$

Clearly, a sufficient condition for obtaining

$$\int t_1 \, d\xi \ \leqslant \ \int t_2 \, d\xi$$

is

$$\int_{\lambda_0}^{\lambda_1} \frac{D_1}{t_1} \, d\xi \ = \ \int_{\lambda_0}^{\lambda_1} \frac{D_2}{t_2} \, d\xi \ = \ \alpha > 0 \tag{34}$$

*) The case of two location parameters is similar.

When the structure is of such nature that D/t does not depend on t, condition (34) is written

(35)
$$\int_{\lambda_0}^{\lambda_1} \frac{D}{t} \, d\lambda = \alpha > 0$$

Hence, optimum can be provided by constructing a family of collapse mechanisms such that the specific dissipation per unit of design variable, integrated with respect to λ, be a positive constant throughout the layout.

When the number of load systems is finite, the integral over λ is replaced by a summation. A system of fixed loads, simultaneous with the movable loads for all values of λ, can also be considered without modification of condition (35).

3.2. Beam Example: One Movable Load

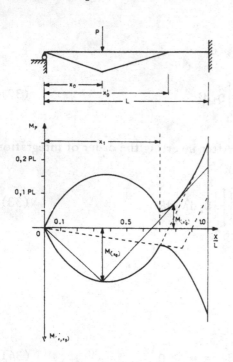

Fig. 5

Consider again the beam of Fig. 1 and load it with a concentrated force P that can occupy any location on the beam.

The kinematical part of the solution, consisting in the choice of a family of collapse mechanisms, will be based on the following consideration. When the load acts at the section with abscissa $x_0 < x_1$ (see Fig. 5), plastic hinges will be assumed to develop at the loaded section x_0 (positive hinge) and some other section with the abscissa $x_0' > x_1$ (negative hinge, Fig. 5). When the positive hinge moves from $x_0 = 0$ to $x_0 = x_1$ with the load, the negative hinge travels from $x_0' = x_1$, to $x_0' = \ell$. Hence, the family of mechanisms exhibits no superposition of hinges in any cross-section. In describing the rate of deflection, it is convenient to regard the

the hinges as narrow regions of the widths dx_0 and dx_0'. In these regions, for each mechanism, the specific description per unit breadth is proportional to the modulus of the rate of curvature which, according to condition (35) (in which no integration over λ is needed here) will be taken as a positive constant that can be arbitrarily set equal to unity. Taking the signs of the hinges into account, we thus have $\dot{K} = 1$ at x_0 and $\dot{K} = -1$ at x_0' over dx_0 and dx_0' respectively. Outside these regions, $\dot{K} = 0$. When the conjugate beam is loaded with these rates of curvature, its equilibrium requires that

$$x_0 . dx_0 = x_0' . dx_0' . \tag{36}$$

Integration of equation (36) gives

$$x_0^2 = x_0'^2 + C . \tag{37}$$

The boundary conditions are

$$x_0' = x_1 \quad \text{for} \quad x_0 = 0$$
$$x_0' = \ell \quad \text{for} \quad x_0 = x_1$$

They furnish

$$C = - x_1^2 \quad \text{and} \quad x_1 = \ell / \sqrt{2} .$$

Hence, we obtain

$$x_0' = \sqrt{x_0^2 + \frac{1}{2} \ell^2} \tag{38}$$

Equation (38) completely defines the family of mechanisms. For $x_0 > x_1 = \ell / \sqrt{2}$, the beam is assumed to remain rigid.

The kinematical part of the solution being completed, we undertake the statical part. We have to express that, upon motion of the load, the bending moment diagram

$$M = M(x, x_0)$$

just touches the curve $M_p = M_p(x)$ in $x = x_0$ and envelopes the branch of the design $M_p(x)$ in $x_1 \leqslant x \leqslant \ell$, being always tangent to $M_p = M_p(x)$ in $x = x_0'$. This requires

(39) $M(x_0, x_0) = M_p(x_0)$

(40) $M(x_0', x_0) = - M_p(x_0')$

(41) $\dfrac{\partial M(x, x_0)}{\partial x_0} = 0$ at $x = x_0'$

As we have three unknown functions, namely $M_p(x_0)$, $M_p(x_0')$ (with $x_0' = x_0'(x_0)$ by (38)) and a redundant $R(x_0)$ (for example the reaction of the simple support), with three equations, we conclude that the problem is well formulated. Practically, we shall solve it without explicitly obtaining $R(x_0)$. The equation of the right-hand branch of the bending moment diagram is:

$$M(x, x_0) = + M_p(x_0) - (x - x_0)\left[P - \frac{M_p(x_0)}{x_0}\right], \quad \text{or}$$

(42) $M(x, x_0) = M_p(x_0) \cdot \dfrac{x}{x_0} - P(x - x_0), \quad 0 \leqslant x_0 \leqslant \ell/\sqrt{2},$

$$x_0 \leqslant x \leqslant \ell$$

Substitution of expression (42) for M in (41) gives

$$x_0' \cdot \frac{d}{dx_0}\left[\frac{M_p(x_0)}{x_0}\right] + P = 0, \quad 0 \leqslant x_0 \leqslant \ell/\sqrt{2}.$$

With the use of (38), integration of this equation results in

(43) $M_p(x_0) = x_0\left[C - \ell n \dfrac{x_0 + \sqrt{x_0^2 + \frac{1}{2}\ell^2}}{\ell}\right], \quad 0 \leqslant x_0 \leqslant \ell/\sqrt{2}$

The integration constant C is determined from the continuity of $M_p(x)$ at $x = x_0$. Indeed, we first substitute the expression (43) for $M_p(x_0)$ in equation (42), and apply equation (40). We obtain

(44) $M_p(x_0') = - x_0'\left[C - \ell n \dfrac{x_0 + \sqrt{x_0^2 + \frac{1}{2}\ell^2}}{\ell}\right] + P(x_0' - x_0)$

Using equation (38) to eliminate x_0 from (44), we have

$$M_p(x_0') = -x_0' \left[c - \ell n \frac{\sqrt{x_0'^2 - \frac{1}{2} \ell^2} + x_0'}{\ell} \right] + P\left(x_0' - \sqrt{x_0'^2 - \frac{1}{2} \ell^2} \right)$$

(45)   , $\ell/\sqrt{2} \leqslant x_0' \leqslant \ell$.

From the condition

$$M_p(x_0) = M_p(x_0') \quad \text{for} \quad x_0 = x_0' = \ell/\sqrt{2}, \quad \text{we obtain}$$

(46) $$c = \frac{1}{2} \left(1 + \ell n \frac{1 + \sqrt{2}}{2} \right)$$

Substitution of expression (46) for C in (43) and (45) finally furnishes the design of Fig. 5.

It remains to show that, for $\ell/\sqrt{2} \leqslant x_0 \leqslant \ell$ this design can remain rigid. This is achieved by finding a statically admissible bending moment diagram for each of these values of x_0. Such a diagram is shown in doted line in Fig. 5(b).

Consider now simultaneous action of the movable load and of any system of fixed loads of section 2.1. Because of equations of equilibrium are linear, simple addition of the bending moment diagrams for the separate loadings gives a bending moment diagram in equilibrium with the simultaneously applied loads. This diagram is statically admissible for a design obtained by simple addition of the respective optimal designs for the separate loads, because M_p is proportional to the design variable. Moreover, the beam obtained in this manner is at collapse with the mechanism corresponding to the case of the movable load alone, because positive hinges of the family of mechanisms cover strictly the region of positive curvature of the collapse mechanism for fixed loads, and similarly for negative hinges and negative curvature. Hence, condition (35) is fulfilled, and we conclude that we have obtained the optimal design for simultaneous fixed and moving loads. This superposition method, already emphasized in section 2.1, holds when the mechanism for fixed loads is "compatible" with the family of mechanisms in the sense given hereabove.

Various applications of the procedure described in section 3.2 can be found in the literature ([11] to [15]).

Of particular interest for what follows is the continuous sandwich beam with two equal spans. The solution is summarized in Fig. 6. Collapse occurs for $0 \leqslant x_0 \leqslant x_1 = \ell/\sqrt{2}$ with a positive hinge at x_0 traveling from $x_0 = 0$ to

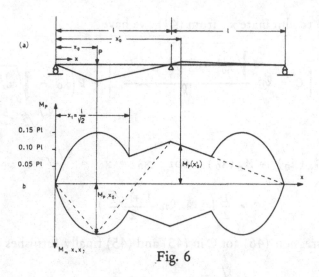

Fig. 6

$x_0 = x_1 = \ell/\sqrt{2}$, and a negative hinge at x_0' traveling from $x_0' = \ell$ to $x_0' = 2\ell - x_1$. It is found that $x_0' = 2\ell - \sqrt{\ell^2 - x_0^2}$. For $x_1 \leqslant x_0 \leqslant 2\ell - x_1$ the beam remains rigid.

The optimal design is given by:

$$M_p = P \cdot \frac{x}{2\ell}\left(\ell + \frac{\ell}{\sqrt{2}} - 2x\right) \qquad \text{for} \qquad 0 \leqslant x \leqslant \ell/\sqrt{2}$$

$$M_p = \frac{P \cdot x}{2}\left(1 - \frac{1}{\sqrt{2}}\right) \qquad \text{for} \qquad \ell/\sqrt{2} \leqslant x \leqslant \ell.$$

3.3. Beam Example: Two Movable Loads

The family of collapse mechanisms of Figure 6 can be used for all systems of (downward acting transversal) loads that make the structure collapse with two plastic hinges. Consider for example the system of two unequal concentrated forces P_1 and P_2 apart from a distance $a < \ell/2$ and acting as shown in Fig. 7 in such a way that the symmetry of the problem with respect to support B is retained. The difference with the problem of Fig. 6 begins with the statical part of the solution, where the question immediately arises to know under which load (P_1 or P_2) is located the positive hinge. Relation (12) gives the answer. Indeed, as $\mathscr{D}$ is fixed for a given mechanism, the right way of associating the system of loads with the mechanism corresponds to the equality sign, and thus gives to the power of the

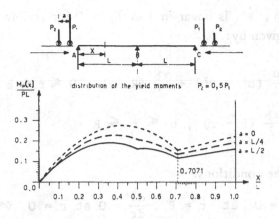

Fig. 7

applied loads the highest value. With this criterion, the problem is solved along the line of section 3.2.

Some other solutions can also be found in the literature [14].

3.4. Circular Plate Example

We turn back to the plate of section 2.2 and load it with a line load of varying radius λ, with $0 \leqslant \lambda \leqslant R$. We assume the load per unit of length to take a constant value P. Hence, the total load $2\pi\lambda P$ is proportional to λ. With this fact in mind, we imagine the following family of collapse mechanisms (Fig.8). For a particular value of λ, say $\lambda = b$, the optimal design is at moment state A (Fig.3) in the central portion $r \leqslant b$ and is at moment states represented by points on AB in the outer region $r > b$. For $\lambda < b$ the whole plate is rigid, while for $\lambda > b$ the central region $r < \lambda$ is rigid and the outer region $r \geqslant \lambda$ is at yield under moment states on AB, with a hinge circle (point A) at $r = \lambda$. The value of b will depend upon the loading, and will be determined later.

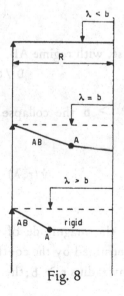

Fig. 8

We denote by v^* the collapse mechanism for $\lambda = b$. It must be such that D^*/t is a constant, say equal to $\sigma_y H\alpha$, for $r \leqslant b$, and because this portion of the plate is at moment state A, this requires (as in section 2.2 eq. (2.4)).

$$\frac{d^2 \dot{v}^*}{dr^2} + \frac{1}{r} \frac{d\dot{v}^*}{dr} = -\alpha \quad , \quad 0 \leqslant r \leqslant b. \tag{47}$$

Also, for $R \geqslant r > b$, $\dot{v}^*$ is linear in r as $\dot{K}_r = 0$ for regime AB. We therefore determine $\dot{v}^*$ to be given by:

(48)
$$\dot{v}^* = \frac{\alpha}{4}(b^2 - r^2) + \frac{\alpha b}{2}(R - b) \ , \ 0 \leqslant r \leqslant b$$

$$\dot{v}^* = \frac{\alpha b}{2}(R - r) \ , \ b \leqslant r \leqslant R$$

where we have used the conditions:

$$\dot{v}^* = 0 \quad \text{at} \quad r = R \ , \ \frac{d\dot{v}^*}{dr} = 0 \text{ at } r = 0, \ \dot{v}^*$$

and $\frac{d\dot{v}^*}{dr}$ continuous at $r = b$.

The mechanisms (48) is such that

(49)
$$\frac{D^*}{t} = \begin{cases} \sigma_y H \alpha & \text{for} \quad 0 \leqslant r \leqslant b \\ \\ \dfrac{\sigma_y}{2} H \alpha \dfrac{b}{r} & \text{for} \quad b < r \leqslant R \end{cases}$$

because, with regime AB,
$$D^*/t = |\dot{K}_\theta| \sigma_y H = \sigma_y H | - \frac{1}{r} \frac{d\dot{v}^*}{dr} | .$$

For $\lambda > b$ the collapse mechanisms $\dot{v}(r, \lambda)$ are of the form

(50)
$$\dot{v}(r, \lambda) = \begin{cases} f(\lambda) \cdot (R - \lambda) & \text{for} \quad r \leqslant \lambda \\ \\ f(\lambda) \cdot (R - r) & \text{for} \quad r \geqslant \lambda \end{cases}$$

where the magnitude of the rate of displacement in the central rigid part is left undetermined by the coefficient $f(\lambda)$

For any radius $r > b$, the optimality condition requires that

(51)
$$\frac{D^*}{t} + \int_b^R \frac{D(\lambda)}{t} d\lambda = \sigma_y H \alpha$$

where D is the specific dissipation derived from (50).

In evaluating D, we must take into account:

   a) the hinge circle, that contributes to D with $M_p \left| \dfrac{d\dot{v}}{dr} \right| = \sigma_y Ht \cdot f(\lambda)$ with $\lambda = r$,   and

   b) plastic flow in regime AB, with $D = \sigma_y Ht \left| - \dfrac{1}{r} \dfrac{d\dot{v}}{dr} \right| = \sigma_y Ht \cdot \dfrac{1}{r} f(\lambda)$.

Hence, condition (51) requires, for $b \leqslant r \leqslant R$,

$$\frac{1}{2} \alpha \frac{b}{r} + f(r) + \frac{1}{r} \int_b^r f(\lambda) d\lambda = \alpha \qquad (52)$$

and it follows that

$$f(\lambda) = \frac{\alpha}{2} . \qquad (53)$$

Equation (53), together with (48) and (50) completely determine the family of collapse mechanisms satisfying the optimality condition, and terminates the kinematical part of the solution.

We now turn to the statical part.

For $\lambda > b$ equilibrium in the outer region of the plate requires:

$$\frac{d}{dr} (r M_r) = \sigma_y H t(r) - \lambda P \quad , \lambda < r \leqslant R \qquad (54)$$

because $M_\theta = \sigma_y Ht(r)$ for regime AB. From (54) we obtain

with $M_r = 0$ at $r = R$,

$$r M_r = - \sigma_y H \int_r^R t(r) dr + \lambda P(R - r) \quad , \lambda < r \leqslant R \qquad (55)$$

and since there is a hinge circle where $M_r = M_\theta$ at $r = \lambda$, we must have

$$\lambda t(\lambda) = - \int_\lambda^R t(r) dr + \frac{\lambda P}{\sigma_y H} (R - \lambda) . \qquad (56)$$

This equation holds for $\lambda > b$ and we deduce that

$$(57) \qquad t(r) = \frac{P}{\sigma_y H} \left[2(R-r) + R \, \ell n \, \frac{r}{R} \right] , \quad b < r \leqslant R.$$

Equations (54) to (57) hold also when $\lambda = b$ (ring load at $r = b$). Further, for $\lambda = b$ the plate is in regime A in the inner zone $r \leqslant b$, and equilibrium, together with continuity of the radial moment M_r at $r = b$, require $t(r)$ to be constant in the central portion:

$$(58) \qquad t(r) = t(b) = \frac{P}{\sigma_y H} \left[2(R-b) + R \, \ell n \, \frac{b}{R} \right] , \quad 0 \leqslant r \leqslant b .$$

The optimal design is therefore known from (57) and (58) when the value of b is known. The value of b is determined from the knowledge that for $\lambda > b$ the region $r < \lambda$ is below yield and for $\lambda < b$ the whole plate is below yield. This is achieved when $t(r)$ is chosen a monotonic decreasing function of r without slope discontinuity, that is

$$\frac{dt}{dr} = 0 \quad at \quad r = b.$$

This last equation immmediately gives $b = R/2$.
With this value of b, it remains:

a) to complete, for $\lambda > b$ the moment field given by (55) and $M_\theta = \sigma_y H t(r)$ by adjoining, in the region $r \leqslant \lambda$ the statically admissible field $M_r = M_\theta = \sigma_y H t(\lambda)$ remembering that $t(r) \geqslant t(\lambda)$ for $r \leqslant \lambda$.

b) to construct statically admissible moment fields for $\lambda < b = R/2$; this is done as follows:

in $r > \lambda$ M_θ is taken positive and less than $\sigma_y H t(r)$, and such that M_r determined by

$$(59) \qquad r M_r = \int_r^R (\lambda P - M_\theta) dr , \quad \lambda \leqslant r \leqslant R$$

is less than $\sigma_y H t(r)$, the feasibility of such a construction being shown by taking M_θ to have the extreme values $\sigma_y H t(r)$ and zero, in turn.

in $r < \lambda$, $M_\theta = M_r = $ constant $\leqslant \sigma_y H t(b)$

It turns out that the fields described in a) and b) can only be constructed for $b = R/2$, and this confirms the choice of this value for b.

3.5. Superposition Theorem for Sandwich Shells

The plate example of section 3.4 demonstrates the appreciable increase of the difficulty of the problem when passing from beams to circular plates. The superposition theorem, already seen in previous sections, will thus prove particularly convenient, when applicable, in plate and shell problems. Consider a sandwich-type shell that has been subjected separately to fixed loads and movable loads, and optimaly designed with $t_f(\xi)$ and $t_m(\xi)$ respectively.

In order to state the superposition theorem we define a condition of "compatibility" between two collapse mechanisms. We say that the collapse mechanism $\vec{v}_1$ is compatible with the collapse mechanism $\vec{v}_2$ if, for any ξ the corresponding strain rates $\dot{\vec{q}}_1$ and $\dot{\vec{q}}_2$ are associated, according to the normality rule, with generalized stress $\vec{Q}_1$ and $\vec{Q}_2$ such that $\dot{\vec{q}}_2$ is associated with $\vec{Q} = \vec{Q}_1 + \vec{Q}_2$. In this definition of compatibility, $\vec{Q}_1$ and $\vec{Q}_2$ are stress states possible for shells with face-sheet thickness t_1 and t_2 respectively, while $\vec{Q}_1 + \vec{Q}_2$ is a stress-state for a shell with face-sheet thickness $t_1 + t_2$. If $\dot{\vec{q}}_1$ and $\dot{\vec{q}}_2$ are both non-zero vectors, the stress states $\vec{Q}_1$ and $\vec{Q}_2$ are at yield and, because of the concavity of the yield locus, compatibility of $\vec{v}_1$ with $\vec{v}_2$ imply that the extremity of vector $\vec{Q}_1 + \vec{Q}_2$ lies o the yield locus for the shell with face-sheet thickness $t_1 + t_2$.

For a strictly convex yield locus, compatibility of $\vec{v}_1$ with $\vec{v}_2$ simply means that $\dot{\vec{q}}_1$ and $\dot{\vec{q}}_2$ are associated with the same point on the yield locus for unit t, and consequently $\vec{v}_2$ is then compatible with $\vec{v}_1$. In general, however, the convex yield locus may contain flats, either plane faces or straight edges, together with vertices and regular surfaces, and comptability of $\vec{v}_1$ with $\vec{v}_2$ does not always imply the converse. Further, if $\dot{\vec{q}}_2$ is zero, that is if $\vec{v}_2$ is a rigid body motion, then any collapse mode $\vec{v}_1$ is compatible with $\vec{v}_2$ and the state $\vec{Q}_1 + \vec{Q}_2$ will lie on or inside the yield locus.

The superposition theorem can now be stated as follows: if the collapse mechanism $\dot{\vec{v}}_f$ employed to determine the optimum design t_f is compatible for each value of λ with the collapse mechanisms $\vec{v}_m(\lambda)$ employed to determine the optimum design t_m then the optimum design for simultaneous action of the fixed and movable loads is obtained by addition: $t = t_1 + t_2$.

The proof is almost immediate. We apply the general procedure for optimality with the collapse mechanisms $\vec{v}_m(\lambda)$ for the design $t = t_1 + t_2$ under the combined loading: fixed and movable loads. From the very choice of the collapse modes, condition (35) is satisfied. We then note that the fields of generalized stresses obtained by superposition of the corresponding fields for fixed and movable loads are in equilibrium with the combined loadings because the equations of equilibrium are linear. Further, these fields of generalized stresses obtained by superposition are at yield for the shell with $t = t_1 + t_2$. Finally we note that the collapse mechanisms $\vec{v}_m(\lambda)$ are associated with these stress fields because $\vec{v}_f$ is compatible with $\vec{v}_m(\lambda)$ for all λ in the considered range.

The superposition theorem above applies to the beam and plate examples treated in the preceding sections. It is also applicable to many optimal designs obtained in the literature (see [13] for discussion).

CHAPTER IV

SECOND GENERALIZATION: STEPWISE VARYING DESIGN UNKNOWN

4.1. Optimality Condition

In many practical problems, the lay-out is divided into m regions A_i in each of which t is restricted to remain constant (frames made of prismatic beams, plates with stepwise varying thickness etc).

The total cost $\mathscr{C} = \int C \, d\xi$ becomes here

$$\mathscr{C} = \sum_i t_i A_i, \qquad i = 1, \ldots, m. \tag{60}$$

We now write equation (32) of section 3.1 in the following way

$$\sum_i \int_{A_i} D_1 \, d\xi \leqslant \sum_i \int_{A_i} D_2 \, d\xi, \tag{61}$$

the two designs being given by the m component vectors $t_{i,1}$ and $t_{i,2}$.

We may re-write (61) into

$$\sum_i t_{i,1} A_i \int_{A_i} \frac{D_1}{t_{i,1} A_i} \, d\xi \leqslant \sum_i t_{i,2} A_i \int_{A_i} \frac{D_2}{t_{i,2} A_i} \, d\xi \tag{62}$$

Note that the integrals are the average dissipation per unit t_i in every region i, and denote them by $\mathscr{D}_i / t_i$. It is then easily seen, by similarity with the previously obtained optimality conditions, that (35) becomes

$$\int_{\lambda_0}^{\lambda_1} \frac{\mathscr{D}_i}{t_i} \, d\lambda = \alpha > 0 \quad \text{for} \quad i = 1, 2, \ldots, m, \tag{63}$$

In beam and frame problems where $t = M_{pi}$ and $\mathscr{D}_i = \sum |\theta| M_{pi}/\ell_i$, summation being extended to all hinges in region i, with length ℓ_i, condition (63) is

$$(64) \qquad \int_{\lambda_0}^{\lambda_1} \frac{\Sigma \, |\theta|}{\ell} \, d\lambda = \alpha > 0 \qquad \begin{array}{l} \text{in all} \\ \text{regions} \end{array}$$

Condition (63) is an extension to movable laods of the well-known condition of Foulkes $\Sigma \, |\theta|/\ell$ = constant.

Despite its outstanding practical importance, and though some analytical solutions exist, we shall not consider here beam or frame examples because these problems are best treated as linear programming problems, a subject that will be developed at length elsewhere*.

We shall accordingly restrict our attention to some circular plates examples.

4.2. Circular Plate Example

Consider again the circular sandwich plate of section 2.2. We now require the unknown thickness t of the face sheets to be constant in circular regions defined by the radii $0 = r_0 < r_1 < r_2 < ... < r_n = R$, the thickness of the region $r_{i-1} < r < r_i$ being denoted by t_i. The values of t_i are to be determined in such a fashion as to minimize the total "cost" φ (face-sheet volume) as given by relation (60), that takes, in the present case, the following form

$$(65) \qquad \mathscr{C} = \sum_{i=1}^{n} \pi \, (r_i^2 - r_{i-1}^2) \, t_i$$

For simply supported circular plates with uniform thickness, subjected to axisymmetrical transversal loading with no sign change, the relevant plastic regime is AB (Fig. 2). This regime will be used in the various annels of the plate. A "hinge circle" (point A) will be assumed in the annel with thickness t at the boundary of that region with the adjacent region of larger t.

For regime AB, the average specific dissipation per unit t is given by

$$(66) \qquad \frac{\mathscr{D}_i}{t_i} = \frac{M_{pi}}{t_i \, A_i} \int_{r_{i-1}}^{r_i} \dot{K}_\theta(r) \, 2\pi \cdot r \cdot dr$$

---

*) G. MAIER "Limit design by linear and integer programming", C.I.S.M. courses on "Structural optimization", October 1974.

According to (22)

$$\dot{K}_\theta = -\frac{1}{r} \cdot \frac{(\dot{v}_i - \dot{v}_{i-1})}{r_i - r_{i-1}} \qquad (67)$$

because regime AB requires $\dot{K}_r = 0$, and hence $\dot{v}(r)$ linear (conical mechanism). Moreover, $\dot{K}_\theta$ must be positive, and thus

$$\dot{v}_i < \dot{v}_{i-1} \qquad (68)$$

(see Fig. 9).
Substitution of (67) in (66) gives

$$\frac{\mathscr{D}_i}{t_i} = \frac{\sigma_y H t_i}{t_i \pi (r_i^2 - r_{i-1}^2)} \int_{r_{i-1}}^{r_i} -\frac{1}{r} \cdot \frac{\dot{v}_i - \dot{v}_{i-1}}{r_i - r_{i-1}} \cdot 2\pi r\, dr \ , \quad \text{or}$$

$$\frac{\mathscr{D}_i}{t_i} = \frac{2\sigma_y H(\dot{v}_{i-1} - \dot{v}_i)}{r_i^2 - r_{i-1}^2} \qquad (69)$$

At a hinge circle with radius r, the (concentrated) dissipation is given by

$$2\pi r |\dot{\varphi}| M_p(r)$$

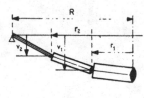

Fig. 9

where $|\dot{\varphi}|$ is the absolute value of the relative rotation of the adjacent parts. For region i, the hinge circle occurs at $r = r_{i-1} + 0$, assuming $t_i < t_{i-1}$ as seems appropriate for the present problem. Hence, the dissipation in the hinge circle is

$$\mathscr{D}_i^c = 2\pi r_{i-1} \cdot \sigma_y H t_i \cdot \left(\frac{\dot{v}_{i-1} - \dot{v}_i}{r_i - r_{i-1}} - \frac{\dot{v}_{i-2} - \dot{v}_{i-1}}{r_{i-1} - r_{i-2}} \right)$$

(see Fig. 9).
Defining

$$\varphi_i = \frac{\dot{v}_{i-1} - \dot{v}_i}{r_i - r_{i-1}} > 0 \ , \qquad \text{and assuming}$$

$0 \leqslant \varphi_1 < \varphi_2 < .. < \varphi_n$ in accordance with (68) and $0 = r_0 < r_1 < .. < r_n = R$, the optimality condition is finally written as

$$\varphi_i \,(r_i - r_{i-1}) \cdot \frac{2\sigma_y H}{r_i^2 - r_{i-1}^2} + \frac{2\sigma_y H \cdot r_{i-1}}{r_i^2 - r_{i-1}^2} \,(\varphi_i - \varphi_{i-1}) = \alpha > 0, \quad \text{or}$$

(70)
$$(\varphi_i\, r_i - \varphi_{i-1} \cdot r_{i-1}) = 2\alpha\sigma_y H (r_i^2 - r_{i-1}^2)$$

Condition (70) is satisfied when $\varphi_i = \beta r_i$ with $\beta = \alpha/2\sigma_y H > 0$, and furnishes the collapse mechanism

(71)
$$\dot{v}_{i-1} = \dot{v}_i + \beta\, r_i\, (r_i - r_{i-1}) \,, \quad \dot{v}_n = 0,$$

compatible with the assumptions made on the variations of $\dot{v}_i, \varphi_i$ and t_i with r_i. For the statical part of the solution, we first integrate equation (27) into

$$\frac{d}{dr}\,(r\,M_r) = M_\theta - \int_0^r pr\,dr$$

With p constant and $M_\theta = M_{pi}$ in region i, we have

(72)
$$\frac{d}{dr}\,(r\,M_r) = M_{pi} - \frac{pr^2}{2}$$

Integration of (72) gives

$$M_{ri} = M_{pi} - \frac{pr^2}{6} + \frac{C}{r}\;.$$

The constant C is determined from the boundary condition

$$M_{ri} = M_{pi} \quad \text{at} \quad r = r_{i-1}\,.$$

We obtain $\quad C = pr_{i-1}^3/6 \quad$ and hence

$$M_{ri}(r) = M_{pi} - \frac{P}{6} \cdot \frac{1}{r} (r^3 - r_{i-1}^3).$$ (73)

The design itself is obtained from the condition $M_{ri}(r_i) = M_{p,i+1}$ that gives

$$M_{p,i+1} = M_{pi} - \frac{P}{6} \frac{1}{r_i} (r_i^3 - r_{i-1}^3)$$ (74)

This formula is used from i = n to i = 0, remembering that $M_{p,n+1} = 0$ ($M_r = 0$ at r = R), a condition that furnishes

$$M_{pn} = \frac{P}{6} \cdot \frac{1}{R} (R^3 - r_{n-1}^3)$$ (75)

It must be emphasized that, in the solution above the radii r_i are considered as given. For every finite number n of these radii, there exist a set of values $(r_1, r_2.., r_n)^*$ that furnishes the plate with the smallest cost. With n = 2, the plate with smallest cost corresponds to $r_1 = 0.8R$ (and $r_2 = R$). Saving with respect to a uniform-thickness sandwich plate is 15% [8] and becomes 18% for n = ∞ (continuously varying t). Similar results have been obtained for a built-in sandwich plate, with the help of an IBM 1130 computer [16]. For uniformly distributed load, savings are: 34% for n = ∞, 12.1% for n = 2, with $r_1 = 0.86R$, and already nearly 20% for n = 4.

CHAPTER V

THIRD GENERALIZATION: ACTION OF BODY-FORCES

5.1. Optimality Condition

Denote by $\vec{F}(\xi)$ the specific body-force, acting on an element of structural layout with coordinate ξ . Consider a collapse mechanism $\vec{v}(\xi)$. If we introduce the modified specific dissipation defined by

$$(76) \qquad\qquad\qquad \Delta = D - \vec{F} \cdot \vec{v}$$

and regard the body-force $\vec{F}$ as independent of P but dependent upon the design unknown t, we can reproduce the reasonings of sections 1.2, 3.1, and 4.1 by simple substitution of Δ for D. For structures where Δ/t does not depend upon t, the optimality condition is

$$(77) \qquad\qquad\qquad \int_{\lambda_0}^{\lambda_1} \frac{\Delta}{t} \, d\lambda = \alpha > 0$$

or

$$(78) \qquad \int_{\lambda_0}^{\lambda_1} \left[\int_{A_i} \frac{\Delta_i}{t_i A_i} \, d\xi \right] d\lambda = \alpha > 0 \quad \text{for} \quad i = 1, 2, \dots, m$$

5.2. Disk Example

Consider a thin circular disk of variable thickness t(r) (Fig. 10), rotating with an angular velocity ω (rad/sec). Let γ be the weight per unit volume of the material. The force of inertia per unit volume has the value $F = \gamma \omega^2 r/g$, where r is the radial coordinate of the considered unit volume and g the gravity acceleration.

Let σ_y be the yield stress of the material and assume that the condition of Tresca applies (Fig. 11).

Consider a collapse mechanism described by the field of radial velocity $\dot{u}$ (r). We know that

$$\dot{\epsilon}_r = \frac{d\dot{u}}{dr} \quad , \quad \dot{\epsilon}_\theta = \frac{\dot{u}}{r} \tag{79}$$

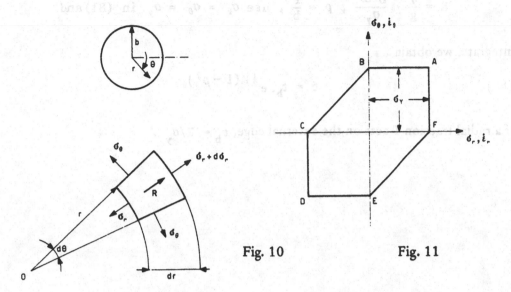

Fig. 10   Fig. 11

and that

$$\frac{\Delta}{t} = \sigma_r \dot{\epsilon}_r + \sigma_\theta \dot{\epsilon}_\theta - \frac{\gamma}{g} \omega^2 r\dot{u} = \alpha > 0 \tag{80}$$

Similarly to the case of circular plates, it can be shown that condition (80) requires the stress regime of point A to prevail throughout the disk. We thus have the condition

$$\frac{\Delta}{t} = \sigma_y \left(\frac{d\dot{u}}{dr} + \frac{\dot{u}}{r} \right) - \frac{\gamma}{g} \omega^2 r\dot{u} = \alpha > 0$$

with

$$\frac{d\dot{u}}{dr} \geqslant 0 \quad \text{and} \quad \frac{\dot{u}}{r} \geqslant 0.$$

The equation of equilibrium is

$$\frac{d}{dr} (r\sigma_r t) - t\sigma_\theta + \frac{\gamma}{g} \omega^2 r^2 = 0 \tag{81}$$

If we let

$$K = \frac{\gamma}{g} \cdot \frac{b^2 \omega^2}{\sigma_y} \ , \ \rho = \frac{r}{b} \ , \ \text{use } \sigma_r = \sigma_\theta = \sigma_y \ \text{ in (81) and}$$

integrate, we obtain

(82) $t = t_b \cdot e^{\frac{1}{2} K (1 - \rho^2)}$

If a radial traction T acts on the external edge, $t_b = T/\sigma_y$.

CHAPTER VI

FOURTH GENERALIZATION: CONVEX COST FUNCTION

6.1. Optimality Condition

Consider, at a generic point ξ of the perfectly plastic structure, its yield locus as depicted in Fig. 12 by curve (a) of a simple two-dimensional example. When ξ varies, the yield curve is allowed to vary its size homothetically, hence retaining its

shape and its position with re-spect to the axes. We call plastic strength of the structure at point ξ the measure $R(\xi)$ of the length of the ray joining 0 to a point of the yield curve in a fixed but otherwise arbitrary direction (for example the length of A if the positive Q_2 direction is chosen). As the specific dis-sipation is

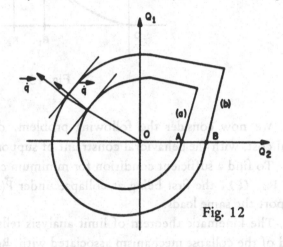

Fig. 12

$$D = \vec{Q} \cdot \vec{q} \quad ,$$

when yield locii are homothetic a proportionality relation is obtained between D and R for given $\vec{q}$:

$$D = k R \quad , \tag{83}$$

where the proportionality factor k depends upon $\vec{q}$ only. By definition, a "design" will be a function $R(\xi)$, R being materialized in any manner (for example by varying the thickness of the structure with one material, or only varying the strength of the material with constant thickness, or both, etc....) provided relation (83) holds.

Assume next that the specific cost of the structure is a convex function of R (see Fig. 13 for example). This property is expressed as follows: for any two values R_1 and R_2 of R, one has

$$C_2 - C_1 \geqslant (R_2 - R_1) \left(\frac{dC}{dR} \right)_{R = R_1} \tag{84}$$

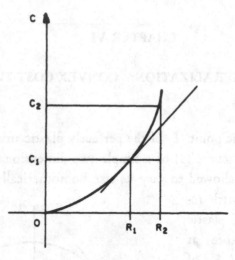

Fig. 13

We now consider the following problem: design the structure for minimum total cost, with the behavioral constraint of supporting assigned loads $\vec{P}(\xi)$*

To find a sufficient condition for minimum cost, compare two designs R_1 (ξ) and R_2 (ξ), the first being at collapse under $\vec{P}(\xi)$ and the second being able to support the same loads.

The kinematic theorem of limit analysis tells us that, if $\vec{v}(\xi)$ is the velocity field of the collapse mechanism associated with R_1 (ξ),

$$(85) \qquad \int \vec{P} \cdot \vec{v} \, d\xi = \int D\left[R_1 (\xi)\right] d\xi$$

$$(86) \qquad \int \vec{P} \cdot \vec{v} \, d\xi \leqq \int D\left[R_2 (\xi)\right] d\xi$$

From comparison of (85) and (86), it follows that

$$(87) \qquad \int \left\{ D\left[R_2 (\xi)\right] - D\left[R_1 (\xi)\right] \right\} d\xi \geqq 0$$

---

*) Concentrated loads will be obtained as uniformly distributed loads acting on very small areas.

Making use of (83), relation (87) is transformed into

$$\int [R_2(\xi) - R_1(\xi)] k [\vec{q}(\xi)] d\xi \geqq 0 \qquad (88)$$

If the collapse mechanism of design $R_1(\xi)$ is such that

$$k [\vec{q}(\xi)] = \left(\frac{dC}{dR}\right)_{R = R(\xi)} \qquad (89)$$

we are allowed, on the basis of relation (84), to substitute $C_2(\xi) - C_1(\xi)$ for $[R_2(\xi) - R_1(\xi)] \cdot k [\vec{q}(\xi)]$ in (88) and we obtain

$$\int [C_2(\xi) - C_1(\xi)] d\xi \geqq 0 \qquad (90)$$

Relation (90) shows that (89) is a sufficient condition for absolute minimum cost.

The mechanical meaning of (89) is as follows: $k [\vec{q}(\xi)]$ is the power of dissipation per unit of strength; dC/dR is the specific cost per unit of strength. Hence, condition (89) states that, in order to obtain a design with absolute minimum cost, it is sufficient to construct a collapse mechanism exhibiting a specific dissipation per unit of strength equal to the specific cost per unit of strength, all over the structure.

Note that both the dissipation per unit of strength and the specific cost per unit of strength are, in general, functions of position on the structure.

We now want to study the case where the description of the loads involves a parameter λ. When λ is assigned a finite number of values, one obtains a multiple loading problem, whereas a moveable set of loads corresponds to continuous variation of λ in some range (λ_0, λ_1). Consider two designs R_1 and R_2 able to support the loads for all values of λ. Assume R_1 is at collapse for some (or all) values of λ. For each such value of λ relations (85), (86), (87), and (88) are valid. Considering all values of λ for which design R_1 is at collapse, $\vec{q}$ becomes a function of ξ and λ and relation (88) must be written

$$\int [R_2(\xi) - R_1(\xi)] k [\vec{q}(\xi,\lambda)] d\xi \geqq 0 \qquad (91)$$

Integrating (91) over λ and assuming that the order of integration may be inverted, we obtain

(92)
$$\int \{[\,R_2\,(\xi)\,-\,R_1\,(\xi)\,]\int_{\lambda_0}^{\lambda_1} k[\,\vec{q}\,(\xi,\lambda)\,]\,d\lambda\}\,d\xi \geqq 0$$

If the collapse mechanisms of design $R_1\,(\xi)$ are such that

(93)
$$\int_{\lambda_0}^{\lambda_1} k[\,\vec{q}\,(\xi,\lambda)\,]\,d\lambda = \left(\frac{dC}{dR}\right)_{R\,=\,R(\xi)}$$

we are allowed, on the basis of (84), to transform (92) into

(94)
$$\int [\,C_2\,(\xi)\,-\,C_1\,(\xi)\,]\,d\xi \geqq 0$$

Hence, (93) is a sufficient condition for absolute minimum cost of $R_1\,(\xi)$. Its mechanical meaning is that the family of collapse mechanisms to use for minimum-cost design should be such that its integrated specific dissipation per unit of strength be at any point of the structure equal to the corresponding specific cost per unit of strength, both being, as a rule, functions of position on the structure.

Condition (93) is readily extended to account for the presence of body forces. Denote by $\vec{v}(\xi)$ the velocity field describing a collapse mechanism, and by $\vec{F}$ the specific body forces (per unit of length of axes, or per unit area of mid-surface). As done in Chapter V, we introduce the modified specific dissipation Δ, defined by

(95)
$$\Delta = D - \vec{F}\cdot\vec{v}$$

We further assume that $\vec{F}$ is proportional to the strength:

(96)
$$\vec{F} = \vec{f}\cdot R$$

where $\vec{f}$ is a given constant vector. Then, from (83), (95), and (96)

(97)
$$\Delta = R[\,k(\vec{q})\,-\,\vec{f}\cdot\vec{v}]$$

Relation (97) is represented in Fig. 14.

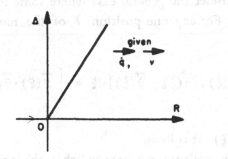

Fig. 14

It is thus seen that simple substitution of Δ for D will give the following sufficient condition for minimum cost:

$$\int_{\lambda_0}^{\lambda_1} \{ k[\,\vec{q}(\xi,\lambda)\,] - \vec{f}\cdot\vec{v} \} d\lambda = \left(\frac{dC}{dR} \right)_{R\,=\,R(\xi)} \tag{98}$$

Hence, in the presence of body forces proportional to the strength, the family of collapse mechanisms to use in order to obtain an absolute minimum cost design should be such that the integrated specific modified dissipation per unit of strength be the same function of position on the structure as the specific cost per unit of strength.

6.2. Discussion of the Sufficient Optimality Condition

In the problem of minimum-weight design of sandwich structures we have the following features:

a) the strength is proportional to the thickness t of the face sheets;
b) the specific cost (volume of face sheets per unit area of midsurface) is also proportional to t:
c) the specific body forces $\vec{F}$ are proportional to the corresponding values of the thickness, and hence proportional to the strength as assumed above.

Indeed, $\vec{f}$ is the weight per unit volume of the face sheets. We thus see that the thickness $t(\xi)$ plays the role of the strength $R(\xi)$, and that, because of (a) and (b), $dC/dR = \alpha$, where α is an arbitrary nonnegative constant. Hence, our general condition (88) reduces to the condition (77) of constant integrated modified dissipation.

6.3. Necessary Condition for Optimality

We immediately consider the general case where some loads are movable and body forces are present. For any one position λ of the movable loads for which collapse occurs, we have

(99)
$$\int \Delta[\, R(\xi), \ \vec{q}(\xi), \ \vec{v}(\xi)\,]\,d\xi \; = \; \int \vec{P}(\xi)\cdot\vec{v}(\xi)\,d\xi$$

whatever the design $R(\xi)$ at collapse.

Considering, for this collapse mechanism, that the applied loads are a one parameter loading with parameter p, and denoting by $\mathscr{P}$ the total power of the loads for unit p, the right-hand side of (99) can be written as

(100)
$$p\cdot\mathscr{P}(\vec{v})$$

If we vary the design by δR, its collapse mechanism varies by $\delta\vec{q}$ and $\delta\vec{v}$ and its load parameter by δp. From (99) and (100) we thus have

(101)
$$\int \left(\frac{\partial\Delta}{\partial R}\cdot\delta R + \frac{\partial\Delta}{\partial\vec{q}}\cdot\delta\vec{q} + \frac{\partial\Delta}{\partial\vec{v}}\cdot\delta\vec{v} \right) d\xi \; = \; \delta p\cdot\mathscr{P}(\vec{v}) + p\cdot\mathscr{P}(\delta\vec{v})$$

But, for any kinematically admissible variations $\delta\vec{v}$ $\delta\vec{q}$ of a true collapse mechanism of a given structure, the kinematic theorem of limit analysis tell us that

(102)
$$\int \left(\frac{\partial\Delta}{\partial\vec{q}}\cdot\delta\vec{q} + \frac{\partial\Delta}{\partial\vec{v}}\cdot\delta\vec{v} \right) d\xi \; = \; p\mathscr{P}(\delta\vec{v})$$

From equation (101) and (102), we conclude that

(103)
$$\int \left(\frac{\partial\Delta}{\partial R}\,\delta R \right) d\xi \; = \; \delta p\,\mathscr{P}(\vec{v})$$

Let us now integrate (103) over the range of λ for which collapse mechanisms occur. We obtain, assuming that the order of integration may be inverted,

(104)
$$\int\!\left\{ \int_{\lambda_0}^{\lambda_1} \left(\frac{\partial\Delta}{\partial R}\,\delta R \right) d\lambda \right\} d\xi \; = \; \int_{\lambda_0}^{\lambda_1} \delta p\cdot\mathscr{P}(\vec{v})\,d\lambda$$

The necessary condition for optimality is

$$\delta \int C[\,R(\xi)\,]\,d\xi = \int \left(\frac{dC}{dR}\,\delta R\right) d\xi = 0 \tag{105}$$

with

$$\delta p = 0 \quad \text{for all } \lambda \tag{106}$$

From (106), the right-hand side of (104) is seen to vanish, and combination of (104) and (105) then gives

$$\int \left\{ \left(\left(\int_{\lambda_0}^{\lambda_1} \frac{\partial \Delta}{\partial R}\,d\lambda - \mu\,\frac{dC}{dR}\right)\delta R \right\} d\xi = 0 \tag{107}$$

or

$$\int_{\lambda_0}^{\lambda_1} \frac{\partial \Delta}{\partial R}\,d\lambda = \mu\,\frac{dC}{dR} \tag{108}$$

Because the mechanisms are defined except for any arbitrary nonnegative scalar coefficient, the Lagrangian multiplier μ can be taken as unity and the necessary condition (108) then coincides with the sufficient condition (98).

Obviously, (108) is established assuming interior extremum and the usual continuity conditions of the calculus of variations. It is worth noting that convexity of the cost function is not used in deriving condition (108).

6.4. Beam Example

a) Consider a built-in beam subjected to uniformly distributed load p and no body-force. Assume $C = 1/2\,\alpha\,M_p^2$.

We have

$$\frac{\partial C}{\partial M_p} = \alpha\,M_p \quad \text{and}$$

$$D = |\dot{K}|M_p \ .$$

Hence, the optimality condition is

$$|\dot{K}| = \alpha\, M_p$$

and the collapse mechanism will be given by the deflected shape of an elastic beam with constant bending rigidity, where $K = \alpha\, M$, with $|M| = M_p$ and sign $K =$ sign M.

This collapse mechanism has contraflexure points at $0.214\,\ell$ of the built-in ends. In these cross-sections $M = 0$ according to the flow law and the continuity of M. We can deduce $M = -\,p\ell^2/12 + px(\ell - x)/2$ and the design

$$M_p = \frac{p\ell^2}{12} \cdot \left| \frac{6x}{\ell}\left(1 - \frac{x}{\ell}\right) - 1 \right|$$

This optimal design is a uniform strength design of a particular statically determinate beam, namely that of Fig. 15b.

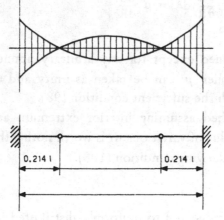

Fig. 15a

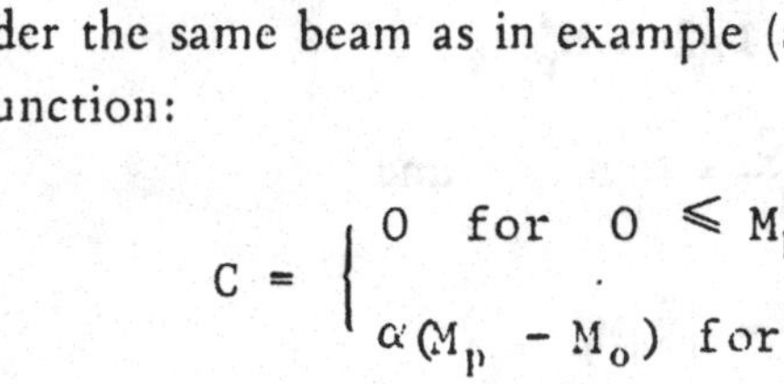

Fig. 15b

0.214 l   0.214 l

b) Consider the same beam as in example (a) but with the following specific cost function:

$$C = \begin{cases} 0 & \text{for } 0 \leqslant M_p \leqslant M_o \\ \alpha\,(M_p - M_o) & \text{for } M_p \geqslant M_o \end{cases}$$

as represented in Fig. 16 we have

$$\frac{\partial C}{\partial M_p} = \begin{cases} 0 & \text{for} \quad 0 \leqslant M_p \leqslant M_o \\ \alpha & \text{for} \quad \quad M_p \geqslant M_o \end{cases}$$

Hence, we must have

$$\dot{K} = 0 \quad \text{for} \quad 0 \leqslant M_p \leqslant M_o$$

$$|\dot{K}| = \alpha \quad \text{for} \quad \quad M_p > M_o$$

If we load the conjugate beam (with free ends), its equilibrium requires a = b. If we denote by X the redundant end moment, we have

$$M = X + \frac{px}{2}(\ell - x)$$

With the collapse mechanism furnished by the optimality condition, the flow law gives

$$M = -M_o \quad \text{at} \quad x = a$$

$$M = +M_o \quad \text{at} \quad x = \frac{\ell - 2a}{2}$$

From these two conditions, we obtain

$$X = \frac{p}{2}\left(a^2 - \frac{a\ell}{2} - \frac{\ell^2}{8}\right) \quad \text{and} \quad \frac{a}{\ell} = \frac{1}{4}\left(1 - \frac{p_o}{p}\right)$$

where $p_o = 16M_o/\ell^2$ is the limit load for $M_p = M_o$ in all cross-sections.

The design procedure is thus as follows:

- if $p \leqslant p_o$, $M_p = M_o$ for $0 \leqslant x \leqslant \ell$

- if $p > p_o$, evaluate $\frac{a}{\ell}$, then X, then M, and take $M_p = |M|$.

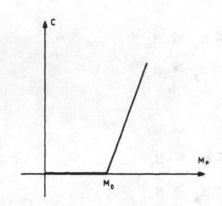

Fig. 16

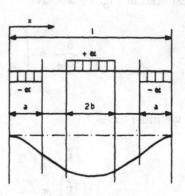

Fig. 17

CHAPTER VII

GENERAL DISCUSSION OF PLASTIC OPTIMAL DESIGN

The preceding chapters were aimed at providing the typical features of the solutions to optimal plastic design problems on the basis of an optimality condition.

Optimal plastic design can also be approached from a basically numerical point of view, expressing directly the minimum condition on the cost functional, without recourse to an optimality condition. With suitable discretization, this formulation is amenable to mathematical programming problems, and more specifically very often to linear programming. ([17], [19] to [24]). Though extremely attractive from a practical point of view, this approach will not be developed here, neither for beams nor for plates, shells and disks, because it is considered at length elsewhere [25].

Even within the limited frame so chosen, many aspects of plastic optimal design have not been touched, because they were not believed to be basic to a proper understanding of the methods. Nevertheless, they deserve being quoted:

— relations of plastic minimum-weight design with elastic uniform- strength design have been pointed out by various authors, and studied in some detail by SAVE [26].

— anisotropic structures can be treated with slight modifications in the procedure. Applications of particular interest are reinforced concrete plates ([18], [27] to [29]) and fiber reinforced shells [30].

— In close connections are the problems of grillages subjected to bending, a field where ROSZVANY and co-workers have made an outstanding contribution ([31] to [34]).

— plastic optimal design for assigned shakedown load is of the highest practical interest when loads are fluctuating. It applies to beams and frames subjected to (slowly) moving loads, and to pressure vessels and tanks. Unfortunately, the theory is still at a rather primitive stage ([36] [37]) except maybe for beams and frames where linear programming gives way to practical solutions.

— as already emphasized in section 2.3, concave cost functions can be encountered in practical problems, the simplest case being that of the rectangular cross-section with variable height for beams in bending. Very little has been done to date in this range ([38], [39]) which offers very wide and promising aspects.

— finally, as will be seen in chapter VIII, plastic optimal design can be framed into a general formulation of optimal design of structures with various behavioral constraints [40].

It is also worth knowing that variational formulations of some optimal design problems have been undertaken [41].

Shape optimization, though not considered here, is an extremely interesting subject, for which a number of solutions have been found, mainly of theoretical interest([42] to [46]), except a recent numerical approach [47] which reveals able to cope with practical problems.

CHAPTER VIII

A GENERAL CRITERION FOR OPTIMAL STRUCTURAL DESIGN

8.1. Design of a Structure

Consider a structure, the general layout of which is given (axes for systems of beams, mid-surface for shells). Denote by ξ an arbitrary point of the layout and by $d\xi$ the corresponding line or surface element. Let q_i , $\dot{q}_i$ and Q_i ($i = 1, ..., n$) be the generalized strains, strain rates and stresses that describe the behavior of the structure, q_i and Q_i being chosen in such a manner that $\Sigma Q_i q_i$ is the work (called specific work because it relates to an element $d\xi$) of the stresses Q_i on the strains q_i, summation being extended over the range $i = 1,2,..., n$.

A design will be defined by m *design functions* $d_1 (\xi)$, $d_2 (\xi)$, ... ,$d_m (\xi)$ such as height or breadth of a rectangular cross section of a beam, thickness of a shell, yield stress of the material, steel reinforcement and orthotropy coefficient in reinforced concrete, modulus of elasticity, moment of inertia etc... In some cases, these functions can be grouped to form functions $R(\xi)$ describing the strength or the rigidity of the structure. For example, in a beam with rectangular cross section with height h, breadth b, yield stress σ_y and Young's modulus E, one can define a plastic strength for bending $R_1 (\xi) = (bh^2/4)\sigma_y$ and a buckling rigidity $R_2 (\xi) = E bh^3/12$.

The unknown functions d or R may sometimes be constituted from known function containing unknown parameters which then become the *design variables*.

8.2. Behavioral Constraints

The structure must be designed in order to exhibit a certain behavior. We consider, for example, the following *behavioral constraints:*

- exhibit a load parameter for plastic collapse not smaller than an assigned value P_ℓ ;
- exhibit an elastic buckling load not smaller than an assigned value P_b;
- exhibit an elastic compliance, measured by the work of the external forces, not larger than an assigned value P_o .
    etc... [48].

All these behavioral constraints will have in common to set a bound on a scalar quantity P which is assumed to be defined by a minimum principle of the type

$$(109) \qquad\qquad P = \min \frac{\int D(q_i)d\xi}{\int H(q_i)d\xi}$$

In equation (109), the functionals D and H are non negative scalar invariants of the field $q_i(\xi)$

with   (110) $D = D \{q_i(\xi), d_1(\xi), \dots d_m(\xi)\}$,

and   (111) $H = H \{q_i(\xi)\}$.

(that is H independent of the values of the design functions, because it is associated to q_i purely by kinematics).

The plastic limit load [49] [50] is given by a kinematic theorem of the type (109), where $q_i(\xi)$ is a kinematically admissible strain-rate field, D the associated specific power of dissipation and P $\int$ H dξ the power of the applied loads. According to the theorem of Rayleigh [51] the elastic buckling load also satisfies (109) where $q_i(\xi)$ is a kinematically admissible strain field, D the specific buckling strain energy and P $\int$ H dξ the work of the loads during the buckling phenomena.

When an upper bound P_o is set on the elastic compliance P, the work of the loads is fixed and equal to P_o. The principle of minimum total potential energy will therefore be expressed by (109) where $\int$ H dξ = 1 and D is the specific elastic strain energy.

When there will be several behavioral constraints associated with the same energetic functionals D, we shall speak of "multiple loading" or "alternative loading".

When the behavioral constraints will be associated with different energetic functionals D_a, D_b... we shall speak of "multi-purpose" structure.

8.3. Cost Function
Let the total cost be defined as

$$C_t = \int C(\xi)d\xi ,$$

where C, the specific cost, will be a single-valued function of the design variables d_1 ..., d_m.

The energetic functions D_a, D_b ... are single-valued functions of q_i and of d_1, ... d_m for every ξ. Hence, for assigned q_i at given ξ, C is a function of the specific "energies" D_a, D_b...

We assume that C $(D_a, D_b, ..., D_u)$ is a *convex function* (Fig. 18), that is

$$C_2 - C_1 \geqslant \sum_i (D_{i2} - D_{i1}) \left(\frac{\partial C}{\partial D_i}\right)_{D_i = D_{i1}} \quad , \quad i = a, b, ... u. \tag{112}$$

where there are u behavioral constraints associated with D_a, D_b,... D_u respectively.

An optimal design must furnish the minimum of C with all constraints satisfied.

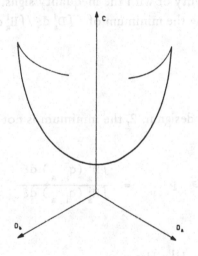

Fig. 18

8.4. Loading

We first consider one set of loads for each behavioral constraint. Next we indicate how the optimality criterion can be extended to multiple loading and even to movable loads. In the discussion of the optimality criterion we also deal with body forces [52].

8.5. Optimality Criterion

Consider first two behavioral constraints

$$(113) \qquad P_a = \min_{q_i} \frac{\int D_a \, d\xi}{\int H_a \, d\xi} \geqslant P_{a,o}$$

$$(114) \qquad P_b = \min_{q_i} \frac{\int D_b \, d\xi}{\int H_b \, d\xi} \geqslant P_{b,o}$$

where $P_{a,o}$ and $P_{b,o}$ are assigned values and the integration extends over the total layout of the structure.

Assume that a design n. 1 satisfies both relations (113) and (114) with the equality sign, and consider a design n. 2 which is simply admissible, that is satisfies (113) and (114) either with the equality or with the inequality signs.

Let $q_{i,a}$ be the field furnishing the minimum of $\int D_a^1 \, d\xi / \int H_a^1 \, d\xi$, that is the value P_a^1 such that $P_a^1 = P_{a,o}$.

With the same field $q_{i,a}$ used with design n. 2, the minimum is not reached and we have

$$(115) \qquad \frac{\int D_a^2 (q_{i,a}) \, d\xi}{\int H_a^2 (q_{i,a}) \, d\xi} \geqslant P_{a,o} = \frac{\int D_a^1 (q_{i,a}) \, d\xi}{\int H_a^1 (q_{i,a}) \, d\xi}$$

But, according to equation (3), $H_a^2 = H_a^1$. Hence

$$(116) \qquad \int D_a^2 (q_{i,a}) \, d\xi \geqslant \int D_a^1 (q_{i,a}) \, d\xi \quad .$$

In a similar manner, we find

$$(117) \qquad \int D_b^2 (q_{i,b}) \, d\xi \geqslant \int D_b^1 (q_{i,b}) \, d\xi \quad .$$

We now multiply both members of (116) by α_a and of (117) by α_b, where α_a and α_b are non negative constants, and add.

We obtain

$$\int [\alpha_a \{ D_a^2 (q_{i,a}) - D_a^1 (q_{i,a}) \} + \alpha_b \{ D_b^2 (q_{i,b}) - D_b^1 (q_{i,b}) \}] \; d\xi \; \geqslant \; 0$$

$$(118)$$

Comparison of (118) and (112) shows that,

$$\left. \begin{aligned} \text{if} \qquad \left(\frac{\partial C}{\partial D_a} \right)_{D_a = D_a^1} = \alpha_a > 0 \\[2em] \text{and} \qquad \left(\frac{\partial C}{\partial D_b} \right)_{D_b = D_b^1} = \alpha_b > 0 \end{aligned} \right\} \qquad (119)$$

then $\int (C_2 - C_1) \; d\xi \; \geqslant \; 0$ and design n. 1 is of absolute minimum cost.

More generally if u behavioral constraints are satisfied with the equality sign by a design, it is sufficient for the design to be optimal that

$$\left(\frac{\partial C}{\partial D_i} \right) = \alpha_i > 0 \; , \quad i = 1, \dots, u . \qquad (120)$$

8.6. Discussion of the Optimality Condition

In its general form (120), the optimality criterion is of little use.

To clarify in particular the number u of behavioral constraints to use, consider the beam of Fig. 19 with rectangular cross section of constant height h and variable

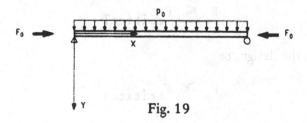

Fig. 19

breadth b. Assume P_0 to be an assigned minimum plastic collapse load and F_0 an assigned minimum elastic buckling load. If P_0 is "large" with respect to F_0, the minimum cost plastic design will generally be rigid enough not to buckle under the action of F_0. Hence, the buckling phenomenon is not relevant to the design and the cost must be regarded as a function of the plastic dissipation D_a only. Conversely, if F_0 is large with respect to P_0 the minimum cost elastic design with a critical load equal to F_0 will not collapse plastically under P_0 Therefore, the cost will have to be expressed as a function of the specific elastic buckling energy D_b only. Finally, if both P_0 and F_0 are relevant, C will be expressed as a function of D_a and D_b and there will be two conditions, that is u = 2 in (120). However, for the problem to be well formulated, the design must be defined by design functions that play a role in both phenomena. In the example of Fig. 19, the design functions are b(x), E(x), $\sigma_y(x)$

Suppose b is given but E and σ_y variable. The two behavioral constraints can be treated separately, furnishing E(x) without influence on the limit load, and $\sigma_y(x)$ without influence on the buckling load. In this case, the behavioral constraints are uncoupled.

On the contrary, if E and σ_y are given and b is the only design function, the two conditions (120) reduce to a single one. Indeed, both D_a and D_b are proportional to b, and hence

$$\frac{\partial C}{\partial D_a} \quad \frac{1}{\beta} \frac{\partial C}{\partial b} \quad \text{and} \quad \frac{\partial C}{\partial D_b} = \frac{1}{\gamma} \frac{\partial C}{\partial b} \quad , \quad \text{where } \beta \text{ and } \gamma \text{ are positive functions of } q_i(\xi).$$

As α_a and α_b in (120) are *arbitrary* positive constants, it is sufficient to let $\beta = \gamma = \partial c/\partial b$ to satisfy the two sufficient conditions of (120).

Finally, if b and E are given, either

$$F_0 \leqslant F_{critical}$$

and p_0 determines the design, or

$$F_0 > F_{critical}$$

and there is no solution. Similarly, if b and σ_y are assigned and only E is free, F_0

determines the design if a solution exists, that is if $P_o < P_\ell$.

It is to be noted that, when one behavioral constraint ceases to be relevant, it is sufficient to make he corresponding constant α in (120) equal to zero.

Clearly, a discussion of the preceding type cannot be made in a general way but only on every specific problem.

8.7. Optimal Design for Assigned Plastic Collapse Load

The general criterion given by the author [52] in 1972 can be recovered in the following way.

Consider first a single system of fixed loads $P(\xi)$ and neglect the body forces. Assume also that a certain "strength" of the structure can be defined as a function $R(d_1, ... d_m)$ of the design variables. For example, in a plate subjected to bending, if h is the thickness and σ_y the yield stress, one can take- $R(\xi) = h^2(\xi) \sigma_y(\xi)$ or $R(\xi) = h^2(\xi)$. If the plate is of the sandwich type $R(\xi) = h(\xi)\sigma_y(\xi)$ or $R(\xi) = h(\xi)$, etc. Assume next that the specific dissipation $D = \Sigma Q_i \dot{q}_i$ is such that

$$\frac{\partial D}{\partial R} = k , \tag{121}$$

where k is a function of $\dot{q}_i$ only, called "specific dissipation per unit of strength".

With a single behavioral constraint, (120) gives

$$\frac{\partial C}{\partial D} = \alpha . \tag{122}$$

But

$$\frac{\partial C}{\partial D} = \frac{\partial C}{\partial R} \cdot \frac{\partial R}{\partial D} . \tag{123}$$

Using (121) and (122) in (123) one obtains

$$\frac{\partial C}{\partial R} = k\alpha , $$

or, more explicitely,

$$\alpha\, k\{\dot{q}_i(\xi)\} = \frac{\partial C\{R(\xi)\}}{\partial R(\xi)} . \tag{124}$$

Condition (124) can be phrased as follows [52] :
"to obtain an optimal plastic design it is sufficient to construct a collapse
mechanism exhibiting a specific dissipation per unit of strength proportional[*] to
the specific cost per unit of strength, all over the structure".

When there are several loading cases, it will be noted, as in § 8.6 that
conditions (120) are simultaneously satisfied with $\alpha_a = \alpha_b = ... = \alpha_u$ when

$$(125) \qquad \alpha \sum_u k = \frac{\partial C}{\partial R} .$$

In the case of a set of movable loads with position such that $\lambda_o < \lambda < \lambda_1$ the
summation over u is replaced by an integration over λ and (125) becomes

$$(126) \qquad \int_{\lambda_0}^{\lambda_1} R \, d\lambda = \frac{\partial C}{\partial R} .$$

In the presence of body forces $\vec{f}$ per unit volume, and if $\vec{v}$ is the velocity field
associated with $\dot{q}_i (\xi)$,(126) becomes [52]

$$(127) \qquad \int_{\lambda_0}^{\lambda_1} \{k - \vec{f} \cdot \vec{v}\} \, d\lambda = \frac{\partial C}{\partial R} .$$

When C is a linear function of R, (127) furnishes the constant dissipation condition
of Drucker and Shield [1] for fixed loads and of Save and Prager [12], and Save and
Shield [13], for movable loads. These conditions have been used in many problems
concerning beams, plates and shells (see, for example, the relevant references in the
books [8], [17], [49], [50] and, more recently, papers [14], [15], [16], [28] and
[53].

Two more remarks are worth making:
a) in the practically interesting case when designing the structure is defined as
   finding t values of R in t portions of the structure with the extent (length or

---

[*]) Note that, with $\alpha = 1$, proportionality reduces to equality.

area) d_i in each of which R remains constant, one has

$$C = \sum_{i=1}^{t} R_i \, d_i \; .$$

It suffices to remark that, in the present situation, D must be regarded as the dissipation in a portion, and k as the (mean) dissipation in a portion per unit of strength, that is $k_i = D_i / R_i$ [16], [54].

b) the specific cost $C(D_a, D_b \dots)$ is a non negative function of non negative variables $D_a, D_b \dots$ If the domain of variation of $D_a, D_b \dots$ is restricted by ϱ conditions of the type $d_i \geqslant d_{io}$ on $d_1, d_2, \dots d_m$, C will be expressed as the *variable part of the cost* in the admissible range for $d_1 \dots d_m$. For values of d_i that would fall outside this admissible range in the case of unrestricted design, one must take $d_i = d_{io}$ and let $\partial C / \partial d_i = 0$.

But

$$\frac{\partial C}{\partial D} = \frac{\partial C}{\partial d_i} \cdot \frac{\partial d_i}{\partial D} = \alpha > 0.$$

Hence, one must have $\partial d_i / \partial D = \infty$, which means that D must not vary with d_i in the regions where an unrestricted design would furnish $d_i < d_{io}$, a condition that will generally imply $D = 0$ in these regions [53] [52].

8.8. Multi-Purpose Structure

Let us consider the example treated by Prager and Shield [54] : a sandwich elastic beam must support an axial load L with a limited axial extension, and alternatively a transverse central concentrated load 2T with a limited mid-span deflection. The minimum volume design under these two behavioral constraints is wanted, the only unknown function being the thickness $t(x)$ of the sheets of the sandwich beam.

The relevant energies are the specific elastic strain energy in axial extension D_a and the specific elastic strain energy in bending D_b. Both are proportional to t, which can thus be regarded as the unknown rigidity $R(\xi)$.

Our general criterion (12) gives:

$$\frac{\partial C}{\partial D_a} = \alpha_a \quad , \qquad \frac{\partial C}{\partial D_b} = \alpha_b \tag{a}$$

As $D_a = D_a(R)$ and $D_b = D_b(R)$, we have

$$\frac{\partial C}{\partial R} = \frac{\partial C}{\partial D_a} \cdot \frac{\partial D_a}{\partial R} + \frac{\partial C}{\partial D_b} \cdot \frac{\partial D_b}{\partial R} = k_a \frac{\partial C}{\partial D_a} + k_b \frac{\partial C}{\partial D_b} \tag{b}$$

from the definition of k.

Using (a) in (b), we obtain

$$\alpha_a k_a + \alpha_b k_b = \frac{\partial C}{\partial R} \;. \tag{c}$$

We now remark that for sandwich structures where C is the thickness of the sheets, $C = \alpha R$, with $\alpha > 0$. Relation (c) thus becomes

$$\alpha_a k_a + \alpha_b k_b = \alpha \tag{d}$$

which is the optimality condition used by Prager and Shield [54].

8.9. Conclusion

The optimality criterion presented above includes most (if not all) previously formulated criterions and is hence believed to enable a unified view of the problem of optimal structural design.

Though difficulties remain in the detailed design procedures, it is hoped that such a synthetic approach will contribute to the access of optimal design to the level of a clear and consistent general theory.

REFERENCES

[1] DRUCKER D.C. and R.T. SHIELD, *Design for Minimum Weight*, in Proc. 9th Int.Cong. Applied Mechanics, Brussels, 1956, Book 5, pp. 212-222.

[2] ONAT E.T., W. SCHUMANN, R.T. SHIELD, *Design of Circular Plates for Minimum- -weight*, Zeit.Ang.Math.Phys. Z.A.M.P., 8: 485, 1957.

[3] FREIBERGER W., B. TEKINALP, *Minimum-weight Design of Circular Plates*, J.Mech-Phys. Solids, 4: 294, 1956.

[4] EASON G., *The Minimum-weight Design of Circular Sandwich Plates*, Zeit.Ang.Math.Phys. Z.A.M.P., 11: 368, Zurich 1960.

[5] PRAGER W., R.T. SHIELD, *Minimum-weight Design of Circular Plates under Arbitrary Loading*, Zeit.Ang.Math.Phys. Z.A.M.P., 10: 421, 1959.

[6] PRAGER W., *Minimum-weight Design of Plates*, De Ingenieur, 67: 141, 1955.

[7] FREIBERGER W, B. TEKINALP, *Minimum-weight Design of Circular Plates*, J.Mech.Phys. Solids, 4: 294, 1956.

[8] HOPKINS H.G., W. PRAGER, *Limits of Economy of Material in Plates*, J. Appl.Mech., 22: 3, 317, 1955.

[9] KONIG J.A., R. RYCHLEWSKI, *Limit Analysis of Circular Plates with Jump Non-homogeneity*, Int.J. Solids and Struct., 2: 3, 493, July 1966.

[10] MROZ Z., *On a Problem of Minimum-weight Design*, Quart.Appl.Math., 19: 3 July 1961.

[11] GROSS O. and W. PRAGER, *Minimum-weight Design for Moving Loads*, Proc. 4th U.S. Nat.Congr.Appl.Mech., 2, 1047, ASME, New York 1962.

[12] SAVE M.A. and W. PRAGER, *Minimum-weight Design of Beams Subjected to Fixed and Moving Loads*, J.Mech.Phys. Solids, Vol. 11, pp. 255 to 267, 1963.

[13] SAVE M.A. and R.T. SHIELD, *Minimum-weight Design of Sandwich Shells Subjected to Fixed and Moving Loads*, Proc. 11th Internat. Congress Appl.Mech., Munich 1964, Springer, Berlin, pp.; 341-349, 1966.

[14] LAMBLIN D.O. and M. SAVE, *Minimum-volume Plastic Design of Beams for Movable Loads*, Meccanica, no. 3, Vol. VI, 1971.

[15] LAMBLIN D.O. *Minimum-weight Plastic Design of Continuous Beams Subjected to One Single Movable Load*, J.Struct.Mech. 1(1), 133-157 (1972).

[16] LAMBLIN D.O. et G. GUERLEMENT, *Dimensionnement de poids minimum de plaques sandwich circulaires constitutuées d'anneaux d'épaisseur constante*, Pub. no. 6 -; Rapport technique R6, 1971. Faculté Polytechnique de Mons.

[17] MASSONNET C. et M. SAVE, *Calcul plastique des constructions*, Vol. 1, Structures *dépendant d'un paramètre*, Centre Belgo-Luxembourgeois d'Information de L'Acier, 47 rue Montoyer, Bruxelles, 2ième édition, 1967.

[18] SAVE M. et C. MASSONNET, *Calcul plastique des constructions*, Vol. 2, *Structures dépendant de plusieurs paramètres*, Centre Belgo-Luxembourgeois d'Information de L'Acier, 47 rue Montoyer, Bruxelles, 2ième édition, 1972.

[19] ZAVELANI-ROSSI A., *Sul progetto ottimale della trave continua ,ercorsa da un treno di carichi,*, Costruzioni Metalliche, n. 3, 187-193, 1970.

[20] SACCHI G. and A. ZAVELANI-ROSSI, *Sulla progettazione ottimale, mediante programmazione lineare, di strutture soggette a carichi fissi e viaggianti*, Istituto Lombardo di Scienza e Lettere, 104, 485-497, 1970.

[21] MAIER G., R. SRINIVASAN and M.A. SAVE,, *On Limit Design of Frames Using Linear Programming*, Proc. of the Int. Symposium on Computer-aided Structural Design, Univ. of Warwick, Coventry, A2.32-A2.59, 1972.

[22] MAIER G., A. ZAVELANI-ROSSI and D. BENEDETTI, *Optimum Plastic Design in Plane Stress Problems: A Finite Element, Linear Programming Approach*, AIMETA Congress Udine 1971, and Int.J.Num. Methods in Eng., Vol. 4, 455, 1972.

[23] SACCHI G., G. MAIER and M. SAVE, *Limit Design of Frames for Movable Loads of Linear Programming*, IUTAM Symposium on Optimization in Structural Design, Warsaw, August 1973.

[24] COHN M.Z., S.K. GHOSH and S.R. PARIMI, *A unified approach to the theory of plastic structures*, Proc. A.S.C.E. J. Eng. Mech. Div., 98, 1133, 1972.

[25] MAIER G., *Limit Design by Linear and Integer Programming*, I.C.M.S. Courses, Saint-Venant Session, October 1974.

[26]   SAVE M.A., *Some Aspects of Minimum-weight Design, in Engineering Plasticity*, Heyman and Leckie ed, Cambridge University Press, pp. 611-626, 1968.

[27]   MORLEY C.T., *The Minimum Reinforcement of Concrete Slabs*, Int. Journ. of Mech. Sciences, Vol. 8, n. 4, pp. 305 319, April 1966.

[28]   SACCHI G. and M. SAVE, *Le problème du poids minimum d'armature des plaques en béton armé*, Mémoires de l'Association Internationele des Ponts et Charpentes, Vol. 29, I , pp. 157-166, 1969.

[29]   MORLEY C.T., *Optimum Reinforcement of Concrete Slabs Element Against Combinations of Moments and Membrane Forces*, Mag. Concrete Research, Vol. 22, n. 72, pp. 155-162, Sept. 1970.

[30]   PAVICIC N.J. and S.C. BATTERMAN, *Optimum Design of Fiber-reinforced Shells of Revolution*, IUTAM Symposium on Optimization in Structural Design, Warsaw, August 1973.

[31]   ROZVANY G.I.N. and S.R. ADIDAM, *Rectangular Grillages of Least Weight*, Proc.-Am.Soc. Civil Eng., Journ. Eng.Mech. Div., pp 1337-1351, December 1972.

[32]   ROZVANY, G.I.N., *Grillages of Maximum Strength and Maximum Stiffness*, Int.J. of Mech. Sci., Vol. 14, pp. 651-666, 1972.

[33]   ROZVANY G.I.N., R.D. HILL and C. GANGADHARAIAH, *Grillages of Least Weight Simply Supported Boundaries*, Int.J. of Mech. Sci., Vol. 15, pp. 665-667, 1973.

[34]   ROZVANY, G.I.N., *Optimal Force Transmission by Flexure: The Present State of Knowledge*, IUTAM Symposium on Optimization Structural Design, Warsaw, August 1973.

[35]   HEYMAN, J., *Minimum-weight of Frames under Shakedown Loading*, Proc.Am.Soc. Civil Eng., Journ. Eng. Mech. Div., pp. 1790-1 to 24, October 1958.

[36]   BORKAUSKAS A., and J. ATKOCIUNAS, *Optimal Design for Cyclic Loaing*, IUTAM Symposium on Optimization in Structural Design, Warsaw, August 1973.

[37]   KONIG, J.A., *On Optimum Shakedown Design*, IUTAM Symposium on Optimization in Structural Design, Warsaw, August 1973.

[38]   ROZVANY G.I.N., *Concave Programming in Structural Optimization*, Int.J.Mech.Sci., Vol. 12, pp 191-142, 1970.

[39]   MARTIN J.B. and A.R.S. PONTER, *The Optimal Design of a Class of Beam Structures for*

a *Non-Convex Cost Function*, Journal de Mécanique, Vol. II, n. 2, pp. 341-360, juin 1972.

[40]   SAVE M.A., *A General Criterion for Optimal Structural Design*, Journ. of Opt. Theory and Appl. (J.O.T.A.), January 1975.

[41]   SACCHI G., *Formulation variationnelle du poids minimal des structures*, Journal de Mécanique, Vol. 10, 4-1, mars 1971.

[42]   MICHELL A.G.M., *The Limits of Economy of Materials in Frame Structures*, Phil.Mag.S., Vol. 8, pp 589-597, 1904.

[43]   HEMP W.S., *Optimum Structures*, Oxford Engineering Sciences Seires, 1973.

[44]   HU T.C. and R.T. SHIELD, *Minimum Volume of Discs*, Zeitschrift fu Angewandete Math. und Physik, (Z.A.M.P.), Vol. 12, Fasc. 5, 1961.

[45]   DORN, W.S., R.E. GOMORY and H.J. GREENBERG, *Automatic Design of Optimal Structures*, Journal de Mécanique, Vol. 3, pp. 25- 51, 1964.

[46]   PRAGER, W., *The Determination of Optimal Layout of a Structure*, I.C.M.S. Courses, Saint-Venant Session, October 1974.

[47]   ZAVELANI, A., G. MAIER and L. BINDA, *Shape Optimization of Plastic Structures by Zero-one Programming*, IUTAM Symposium on Optimization in Structural Design, Warsaw, August 1973.

[48]   PRAGER W., *Conditions for Structural Optimality*, Computers and Structures, Vol. 2, pp. 833-840. Pergamon Press 1972.

[49]   PRAGER W., *An Introduction to Plasticity*, Addison-Wesley Pub. Comp. Inc. USA, 1959.

[50]   SAVE M. and C.H. MASSONNET, *Plastic Analysis and Design of Plates, Shells and Disks.*, North Holland Pub. Comp. Amsterdam, 1972.

[51]   FLUGGE W., *Handbook of Engineering Mechanics,*, McGraw—hill Book Comp., Inc. New York, 1962, Ch. 44.

[52]   SAVE M., *A Unified Formulation of the Theory of Optimal Plastic Design with Convex Cost Function*, J.Struct.Mech., 1(2), pp. 267-276, 1972.

[53]   MARCAL P.V. and W. PRAGER, *A Method of Optimal Design*, J.Mecan. 3(4), 509-530 (1964).

[54]   PRAGER W. and R.T. SHIELD, *Optimal Design of Multi-purpose Structures*, Int.J. Structures, 1968, Vol. 4, pp. 469 to 475.

Printed in the United States
By Bookmasters